A-Z
CELL BIOLOGY

A-Z
CELL BIOLOGY

Dewasish Choudhary

CENTRUM PRESS
NEW DELHI-110002 (INDIA)

CENTRUM PRESS
H.O.: 4360/4, Ansari Road, Daryaganj,
New Delhi-110 002 (India)
Ph.: 23278000, 23261597

B.O.: No. 1015, Ist Main Road, BSK IIIrd Stage
IIIrd Phase, IIIrd Block,
Bangalore - 560 085 (India)
Tel.: 080-41723429
Visit us at: www.centrumpress.com

A-Z Cell Biology

First Edition, 2009

ISBN 978-93-80106-11-3

PRINTED IN INDIA

Printed at Salasar Imaging Systems, Delhi-110035 (India)

Contents

Preface

Cell Biology is an important link between almost all Biological Sciences, and is assuming more and more importance in fields such as *Medicine and Agriculture*. It is an academic discipline that studies cells – their physiological properties, their structure, the organelles they contain, interactions with their environment, their life cycle, division and death.

This is done both on a microscopic and molecular level. Cell biology research emcompasses both the great diversity of single-celled organisms like bacteria and protozoa, as well as the many specialized cells in multicellular organisms like humans.

Cell Biology has began fulfilling its long anticipated role of linking genetics and embryology and many questions of cytology that had lain dormant for decades have been taken up with new molecular tools and strategies.

Cell Biology deals with the Cell structure. All cells consist of Protoplasm contained within a Surface Membrane, Plasma Membrane or Cell Membrane. They contain double-stranded DNA as *Genetic Material*. Although there are many different types of cells which, through the process of differentiation, become specialized for particular roles.

Author

Chapter 1

Introduction to Cell Biology

Far away in the distant past, when the primeval seas bubbled and steamed, when ultraviolet light of great intensity streamed onto the surface of the earth, molecules of polypeptides, of long-chain fatty acids, steroids of various sorts and purines and pyrimidines ultimately to form part of the nucleic acids, went through the seething pangs of synthesis. Both purines and pyrimidines have recently been found in the interior of meteorites —thus they are not unique to this world and indicate that nucleic acid synthesis may be occurring elsewhere in space.

These compounds, floating free in a warm soup, performed no organized activity. They reacted indiscriminately with each other, blindly, with no form or purpose. Then, groups of molecules responding to electrical and surfaceforming forces became orientated in a way which produced a membrane.

Presumably these membranes which were forming on the surface of the water were at first simply flat structures, but "One day" through inter-reaction of the various molecules a membrane formed a little vesicle. Many other vesicles became formed and here we had for the first time the potentialities for the development of living structures, for the membrane cut off the contents of the bag from the environment. What could have been the composition of such membranes? Perhaps the amino acids were formed into proteins at the time that this happened. Proteins, in general, have a remarkable property of spreading in a very thin film upon the surface of water—even when they are themselves in solution they are capable of forming these thin films. Compounds such as steroids and fats, i.e., compounds which have a preponderance of hydrocarbon groupings which makes them only slightly soluble in water, also have the ability to spread on surfaces and to have a particular orientation so that their polar groups will, in general, be in contact with water or with other polar groups and those parts which have hydrocarbon groups will be out of contact with the water. Furthermore, lipoprotein films which combine both these two types of compounds, can even be formed artificially, for instance cholesterol and the protein gliadin have been used to produce such a film. These films have considerable elasticity and a good deal of strength.

Within the little vesicles we have mentioned, the reorganization of the molecules went on and eventually a vesicle formed within the parent vesicle which formed a structure similar to the nucleus. Combinations between proteins and nucleic acids formed a basis for inheritance of some of the characters of the cell in a way which we do not yet understand fully. By a slow and gradual process of evolution a complicated structure we now know to be a cell was built up and from it developed the multicellular organisms, animal and plant. At the stage when the multicellular organisms were evolving, it is not even certain that the cell was completely developed in the way we know it at the moment, it probably had a much simpler character than present day cells.

The bags or vesicles contained proteins, fats, phospholipids, carbohydrates, minerals, enzymes, vitamins and water. Now if one normally shook such components up in a little bag not much would happen, although no doubt there would be some linkup of some of the molecules to form structures. For example, if collagen is disaggregated with citrate and then is left to stand after the citrate had been removed, many of these groups of molecules reform into collagen units or even collagen fibers. Something of this sort could occur in the type of vesicle we have described if the right molecules got together. On the other hand, in the beginning very little activity characteristic of the living cell would develop. In the presence of their substrates, for instance, many of the enzymes present would have some action, but for some of the enzymes the pH would be wrong and in such a mixture there would probably be only one uniform pH and the temperature would not be optimal. There would be no mechanism for removal of reaction products from the site of their formation and the reactions would be rapidly saturated with them and would grind to a stop.

The secret of the complex activity of the living cell lies in the isolation or partial isolation or even the timed isolation of its various activities. This is achieved by the presence of various membranes. The nuclear membrane guards the activity of the nucleus and the Golgi apparatus and the mitochondria have membranes which permit them to carry out activities in partial but not complete seclusion.

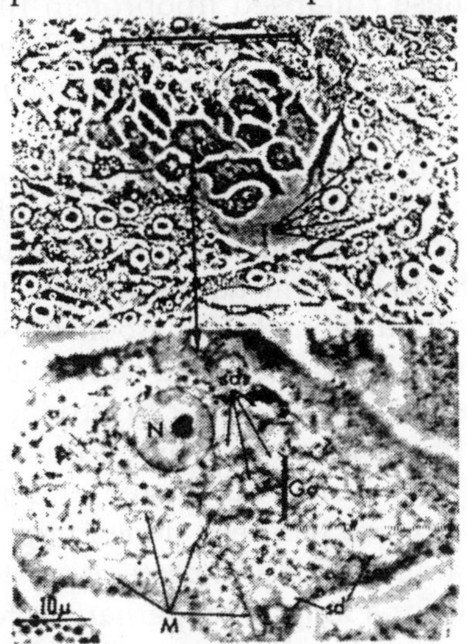

Fig. The Living Cells Under the Optical Microscope.

It is of interest that natural, fresh water frequently contains films and these are probably derived, according to Goldacre, from decomposed biological material. They appear to contain protein or lipoprotein and in many ways possibly resemble the type of film just described from which cells were probably originally formed. It is of further interest that films such is these, upon collapsing, can spontaneously form vesicles or cylinders, e.g., the wind may cause a collapse of the film. In the case of water passing under a floating barrier, for example, when a stream disappears underground, the surface film is carried toward the point of entry and according to Goldacre, may be concentrated.

Compression of this film causes a wrinkling and if the wrinkles become sufficiently pinched the walls finally touch underneath the wrinkles and complete cylinders may be detached which separate from the rest of the film. These cylinders themselves can form vesicles which would resemble the primeval vesicles previously mentioned; some may contain air and some the underlying water. Sometimes instead of cylinders, fibers may be produced. It is of interest that the properties of these collapsed lipoprotein films (which can actually be produced in the laboratory) have, according to Goldacre, many similarities to those of the membranes of living cells.

Proteins and other macromolecules, according to Brownell, seem to be able to form in nature without any special difficulty. Presumably small organic molecules could form quite easily under the circumstances in which the earth existed at one time. Carbon monoxide, carbon dioxide, hydrogen, nitrogen and other gases including some inorganic catalysts such as various metals and ions can produce small organic chemicals which, once formed may go through a process of polymerization to build up larger and more complex molecules. This, in fact, has been done in the laboratory. It is very likely that amino acids and proteins existed in the soupy seas of the earth for a long time, perhaps before membranes formed; on the other hand, it is possible that as the former substances were produced many of them became organized into surface films which became membranes and subsequently divided up into vesicles.

As mentioned earlier, artificial membranes can easily be formed and mixtures of fatty substances, alkalis, proteins and inorganic salts are very suitable for this purpose. Hereira has even formed in this way, what he describes as artificial amoebae which were persuaded to produce types of amoeboid movement. But, as Goldacre points out, such movements are uncontrolled and what is so important in cells is the fact that there is a very considerable degree of control by the cell over itself. Nevertheless the studies of Hereira help us perhaps to understand how objects comparable to cells, in other words, isolated vesicles containing macromolecules could have been produced in the first place.

Goldacre himself points out that the essential step in the evolution of the cell was the formation around a chemical system of what he describes as a relatively impermeable envelope in the beginning. But, as he also points out, what was presumably a relatively simple structure at first must have undergone a considerable amount of further development and membranes of present day cells are relatively complicated structures which take quite an active part in regulating the compounds and ions that go in and out of the cell. For this purpose they seem to have absorbed, adsorbed, or incorporated enzymes in some way onto or into their surfaces and these enzymes play a part in the process of active transport.

Chapter 2

The Structure of the Cell Membrane

Let us trace briefly the origin of our present knowledge of the structure of cell membranes. Overton in 1895, was originally responsible for the idea that lipids were important in cell membrane structure. He made this suggestion because of the relative ease with which lipid soluble substances penetrated the cells. At that time there was no clear idea as to what types of lipid molecules were present or as to how they were arranged.

In 1917, Langmuir and other workers made deductions about the arrangement of molecules in molecular films spread on water surfaces and these have been described as being of fundamental importance to the modern concepts of the cell membrane structure. Langmuir suggested (as mentioned in the introduction) that lipids at an interface of air and water arrange themselves in a monolayer with the polar ends of the molecules directed toward the water interface and the nonpolar ends at the air interface. In 1925, Gorter and Grendell found that the total amount of lipid they could extract from a red cell.

Ghost when spread in a monolayer occupied about twice the area of the membranes of the red cells. They suggested, therefore, that the lipid was arranged in the membrane as a bimolecular leaflet in which the hydrophilic polar groups were at the inner and outer surfaces and the hydrophobic carbon chains were directed toward each other in the interior of the membrane. This was a simple and attractive model. It is difficult to imagine however how such a simple structure could account for all the various functions of the cell membrane.

Dr. G. D. Robertson has described the next stages of the development of our knowledge of the cell membrane. Briefly these are as follows: Harvey and Shapiro in the 1930's and also Cole during the same period studied the surface tension of starfish eggs and came to the conclusion that, at the ranges of pH's which apparently occurred inside cells, most pure lipoid substances gave surface tension values of the order of 5 dynes/cm.^{2} or more. But intracellular oil drops were found to have a surface tension of only 0.2 dyne/cm.^{2} The surface tension of whole starfish eggs as estimated by the force required to compress them, is even lower than this.

Thus, if only a purely lipid/water interface is concerned, these low values for the living cell cannot be explained. Danielli and Harvey demonstrated that there was some substance on the cell membrane which was responsible for such low values. This substance was a surface active agent of a cytoplasmic nature and appeared to be protein. Apparently in natural membranes all lipid polar substances were covered with at least a monolayer of protein. Danielli and Davson eventually proposed that cell membranes in general had the same sort of structure and that this consisted of one or more bimolecular leaflets of lipid with each polar surface having on it a monolayer of protein.

In 1936, Schmitt, using polarization microscopy, concluded that at least in red cell membranes the lipid molecules which were present had their carbon chains radially oriented. Thus we can picture the cell membrane as being composed of two layers of lipid molecules radially arranged (at right angles to the surface of the membrane) and a monolayer of protein applied to both the inner and outer surfaces of the membrane, with the long axes of the molecules lying parallel to the surface.

It had been suggested in the past that the cell membrane contained either pure lipid or pure protein, with pores of dimensions about the size of a large molecule, or that it was in the form of a mosaic containing either pure lipid or pure protein in various areas. However, this seems unlikely now except in some special cases. It is of interest that a common structure for the cell membrane exists in cells as diverse as erythrocytes, axons of nerve cells, muscle fibers, leucocytes of the blood, yeast cells, algal cells, the cells of higher plants and the ova of echinoderms, e.g., sea urchins and starfish. It seems very likely, however, that there are variations to some extent between the membranes of the various cells; nevertheless it is highly probable that a general structural pattern does exist for all of them.

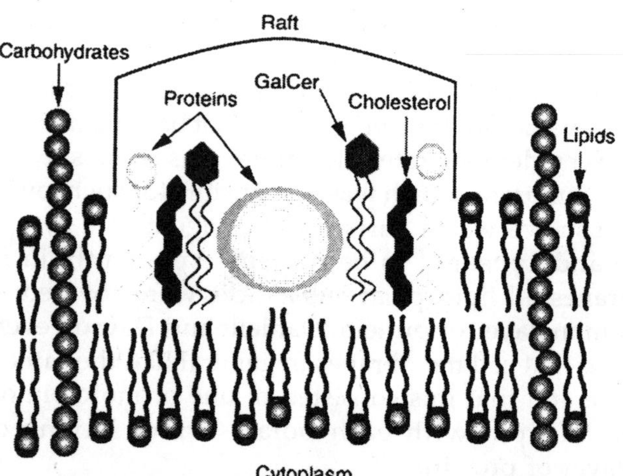

Fig. The Structure of the Cell Membrane

A typical example of variation was demonstrated by Mudd and Mudd who showed that erythrocytes are preferentially wetted by oil and leucocytes by water; this indicates that molecules at the surface of the membranes differ.

Studies with the electron-microscope have demonstrated that those membranes which have been studied have a thickness of the order of about 80 A (10,000 A = lã) so that they are about thick. Dr. Keith Porter has pointed out that the first electron-microscope photographs demonstrated the cell membrane as a solid structure, but that, as the technique improved and better resolution was obtained, the membrane was seen to be composed of two thin, very dense lines with a space between them. The lines measured about 25 A each and the space between them measured about 30 A across. It is tempting to suggest that the two outer lines represent the protein parts of the membrane and the central space the phospholipid portion, but this has yet to be proven. This electron-microscope pattern of structure and size is general in cell membranes—it is present, for example, in the nuclear membrane, the membranes of the endoplasmic reticulum and those of the mitochondria. In due course no doubt the electron-microscope will shed further light on the nature of the membranes and possibly the differences between the many types of membranes in various cells, but there are limitations to what it can do at the present moment.

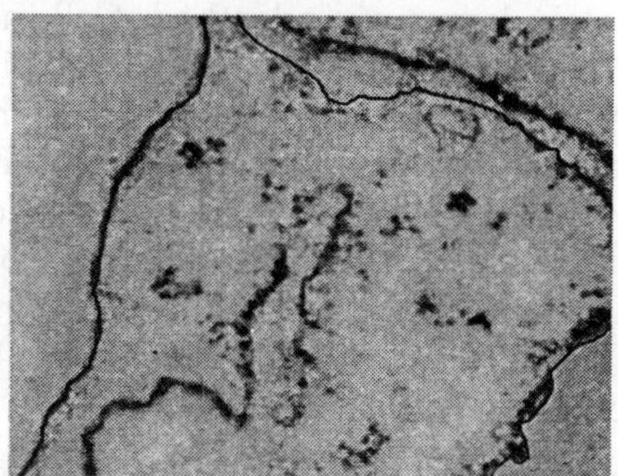

Fig. The Cell Membrane as Demonstrated in the Electron Microscope.

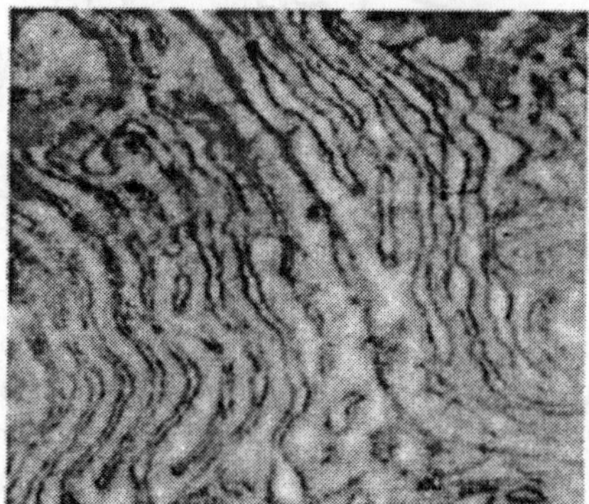

Fig. The Cell Membranes at the Junction of Two Acinar Pancreatic Cells

Some of the ideas of the nature of the cell membrane have been derived from a study of its fundamental properties. First of all it is known that lipid-soluble substances have a preferential permeability across cell membranes and this suggests that there must be in the membrane a continuous layer which is composed of phosphatides, steroids or fats or combinations of those substances. Also cell membranes have a high electrical resistance which, Danielli points out, is further evidence that there must be a continuous layer of lipids. Another property is the existence of a low surface tension at the surface of the cell membranes.

We have mentioned earlier how Danielli has suggested that this property indicates that on the surfaces of the membrane protein layers are adsorbed. Cell membranes are disrupted by digitonin which has a special ability to form a complex with cholesterol and this suggests that this latter compound may play a part in the structure of the membrane.

Analysis of red cell ghosts has in fact demonstrated that a relatively large amount of cholesterol is present in the membrane. An interesting theory concerning the structure of the red cell membrane was proposed by Winkler and Bungenburg de Jong in 1941.

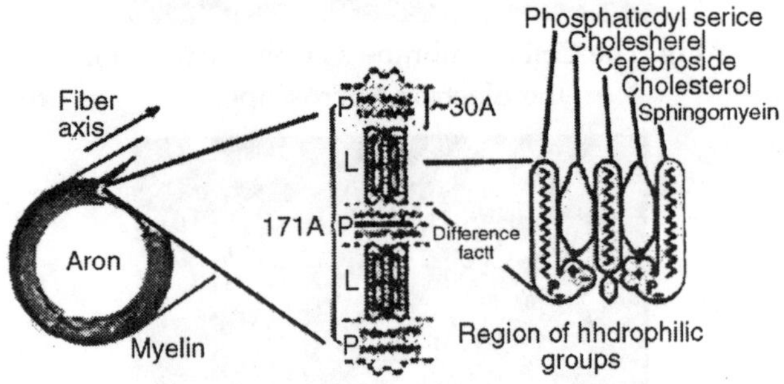

Fig. Molecular Structure of the Myelin Sheath

This theory gives the free cholesterol the role of stabilizer for the charged phospholipid molecules thus enabling them to give form to the membrane. This concept has been elaborated by Frey-Wyssling in 1953, whose studies on submicroscopic morphology and its relation to the chemical constituents of the cell are of fundamental importance. It is of interest to notice that in the two issues where cholesterol has been assigned the major role of structural component, it is present mainly if not entirely in the free unesterified form. There is one striking difference however in the nature of the cholesterol in the myelin sheath and the red cell.

In the myelin sheath it is apparently stable and once deposited under normal conditions remains static for life. In the red blood cell there is a steady turnover and exchange with plasma cholesterol. In general, in tissues where there is active metabolism of cholesterol, esterified cholesterol is found, the amount varying from tissue to tissue.

One should, perhaps, at this point say a word about the composition of the myelin sheath since this is in a sense an extended membrane. Finean, from his studies on the x-ray diffraction pattern of myelin, suggested that it was probably composed of alternate layers of lipid and nonlipid, the nonlipid probably being protein. The lipid molecules appear to be curled or tilted so that their full length is not extended and there is probably a stable complex formed between the cholesterol molecule and the larger phospholipid molecules. It is postulated that the free hydroxyl group of the cholesterol is important in binding the molecule by associating with the polar end of the lipid chain which it curls round. The hydrocarbon part of the cholesterol molecule is bound to the phospholipid by Van der Waal's forces. Cholesterol has a high dielectric constant and because of this it might act as an insulating agent.

THE PERMEABILITY OF CELL MEMBRANES

Danielli has discussed on several occasions the various problems associated with the permeability of natural membranes and the account which follows is taken largely from his writings.

PENETRATION OF COMPOUNDS

Danielli points out that there are three sites of resistance to free diffusion in cell membranes:

- The membrane/water interface for diffusion into the membrane;
- The membrane/water interface for diffusion from the membrane into the water;
- The interior of the membrane.

Every molecule which requires to pass into the interior of the cell has to get through these three sites of resistance. A molecule which has polar groups (-OH groups), as opposed, for instance, to nonpolar groups such as methylene, forms at least one hydrogen bond with water for each polar group; all these hydrogen bonds must be fractured simultaneously if the molecule is to penetrate even into the lipoid membrane. Glycerol, having three OH groups, must acquire sufficient kinetic energy to break three hydrogen bonds simultaneously before it can penetrate into the membrane. This involves a large amount of energy, consequently resistance 1 (diffusion from water to membrane) is so high for glycerol that resistances 2 and 3 are reduced to insignificance.

On the other hand, a molecule such as methyl alcohol which has only one OH group and one CH_3 group penetrates easily into the membrane and can also pass easily out of the lipoid layer into water and presumably ethyl alcohol (C_2H_5OH) can function similarly. Thus we find that for such molecules the rate of diffusion across the interfaces is very large compared with rate of diffusion across the membrane. The interior of the membrane in the case of these compounds is the most important factor controlling penetration. Differing from these examples are molecules which are predominantly hydrocarbon in nature such as carotene ($C_{40}H_{56}$).

With a molecule such as this, resistance I is insignificant; the molecule has no polar groups and thus even resistance 3 is not enormous; however, resistance 2 is very large indeed, because the hydrocarbon groups are hydrophobic and a considerable amount of kinetic energy is required to transfer CH_2 groups from lipid into water. When many such groups are present they must all be transferred simultaneously from the lipoid layer into water, since otherwise the molecule remains substantially part of the lipoid layer and cannot diffuse away into the aqueous phase.With these examples in mind, Danielli classified penetrating molecules into four groups.

- Molecules with few polar and few nonpolar groups —for these resistance 3 is most important and they can penetrate comparatively rapidly e.g., oxygen and methyl alcohol.
- Molecules having a predominantly polar character-for these resistance 1 is most important and penetration is slow, included are glycerols, sugar and glycogen.
- Molecules having few polar and many nonpolar groups—resistance 2 is most important, penetration is slow e.g., carotene, vitamin A and fat.
- Molecules having many polar and many nonpolar groups—resistances 1 anti 2 are both important and penetration is slow e.g., polyhydroxylic bile acids, the glucuronide of oestrin and proteins.

Danielli has pointed out that these principles of membrane permeability have a distinct physiological significance. For instance, oxygen is required in large amounts by the cell and carbon dioxide must be disposed of rapidly and both of these substances can penetrate the cell membrane rapidly. On the other hand, the first products of glucose utilization, for example, by muscle cells are glycerol derivatives, these are valuable and the cell membrane does not let them get through too easily and so they escape only slowly. During sustained work lactic acid is formed—this would be toxic if it accumulated. The cell membrane is relatively permeable to lactic acid which thus can escape into the blood and this permits violent exercise to be maintained for a much longer period than would be the case if the cell membrane was impermeable to lactic acid.

Amino acids penetrate all membranes moderately well but protein penetrates badly and, therefore, amino acids are stored as protein. Fatty acids will penetrate moderately well, but neutral fat very poorly, therefore fatty acids are stored as neutral fat. Glucose which can get in and out of cells fairly rapidly is polymerized in liver cells to form glycogen which passes the cell membrane only with great difficulty. Each of these three latter products, to which the membranes are impermeable, is the results of polymerization of simpler compounds.

It is of interest that, when compounds which are diffusing through a cell membrane from outside to inside are being incorporated into some compound within the cell, they appear to penetrate the membrane more easily. A typical example of this can be drawn from amino acids. If they are being actively incorporated into proteins, the rate of diffusion through the membrane will be much more rapid than if they were not being incorporated. Presumably, if they are being built into proteins, they cease to accumulate and so cease to build up a concentration gradient against the passage of further amino acids.

Detoxification mechanisms also take advantage of the principles of the cell membrane permeability. For instance, toxic substances which might penetrate the cell membrane become conjugated with amino acids, sulfuric acid, or glucuronic acid. Thus toxic cell-penetrating substances such as bromobenzene or menthol are converted into new molecules which penetrate the cells with great difficulty and once in the bloodstream tend to be filtered off by the glomeruli of the kidneys and cannot be reabsorbed from the urine and returned to the bloodstream by the kidney tubules.

Penetration of Ions Through Cell Membranes

Penetration of ions through cell membranes is of special interest because for instance, Na^+ions are more concentrated outside the cell (in the body fluids) and K+ions are more concentrated inside the cell, yet we know that K ions will pass into the cell and Na ions will pass out. Both these passages thus take place against a concentration gradient and this can only occur by the application of energy. It is claimed that oxidative enzymes and phosphatases play a part in the movement of ions in and out of membranes; a process which is known as "active transport."

Substances which interfere with the activity of these enzymes have been shown in many cases to interfere with the movement in or out of the cell of both sodium and potassium. There are certain specialized cell membranes, e.g., the cell membrane of the intestinal cells, in which the distal edge of the cell is folded to produce a larger number of extremely small fingerlike processes called microvilli; these microvilli are covered with a typical cell membrane. There are a large number of dephosphorylating enzymes localized around these microvilli. Their precise function in this position is not known, but it is almost certain, however, that they play a part in active transport by utilizing the energy derived from the hydrolysis of high-energy phosphates.

It is obvious from what we have said, therefore, that the cell membrane, in general, is a very selective structure anti exercises a good deal of control over the cell by deciding What goes in or out and by enabling the cell to be isolated from its environment. In this way it permits labour to go on in the cell which may be quite different from that going on in a neighboring cell, or in the body fluids surrounding the cell. Here then is the first division of labour in the cell to be described by us.

However the story is not yet ended. In 1952-54, Danielli proposed the structure of the cell membrane which explained the way in which large protein molecules in the cell membrane could permit the passage of ions through the membrane. Danielli conceived of a long protein molecule extending right through the membrane and passing outside it. This end became attached to an ion and then by contraction pulled the latter through into the interior of the membrane.

According to Lundgard and Hodgkin, a cytochrome oxidase energy-producing system is involved in the penetration of ions. They claim that their ion moving system requires adenosine triphosphatase (ATPase), creatine phosphatase and cytochrome oxidase. Conway has produced a redoxpump hypothesis that attempts to explain the penetration of ions through cell membranes by a mechanism which involves the successive oxidation and reduction of a compound in the membranes; possibly a phospholipid provides the energy for the movement of ions in this way. Phosphatidic acid is now thought to be concerned with ion movements through membranes.

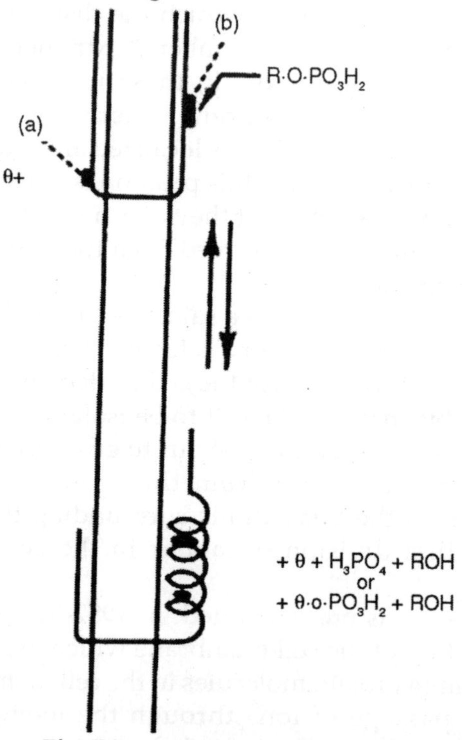

Fig. Diagram of Danielli's Theory of Secretion by a Contractile Protein

The possible relationship of phosphatases to cell membrane permeability has also been mentioned in connection with the microvilli (brush borders) of intestinal epithelium. The passage of glucose across the mebranes of gut cells could be explained by a mechanism involving these enzymes. It is of interest that in yeast cells too there is some evidence that phosphatases are localized on the membrane and that these enzymes are also involved in the permeability not only to glucose but also to phosphates. Phosphatases are also present on the brush borders of the kidney tubule cells where a good deal of absorption takes place. More recent work has indicated that ribonuclease may play a part in the penetration of cell membranes by ions. Lansing and Rosenthal and Tenardo, for instance, showed that when ribonuclease was added from the outside to certain cells, it modified their permeability to ions; Brachet and Leduc found exactly the same thing for amphibian eggs. According to Brachet "amphibian eggs swell very much when they are immersed in a ribonuclease solution and since cell membranes often give strong cytochemical tests for ribonucleic acid it might well be that the integrity of this nucleic acid is of importance for normal permeability."

Danielli made a suggestion that the cell membrane consisted, as suggested before, of bimolecular leaflets of lipoid with protein molecules stretched both inside and outside the membrane, but that it also contains a series of pores which he calls polar pores. In other words, groups of protein molecules are oriented radially with their polar groups directed toward the interior of the pore and such molecules are thus able to control the penetration of the compounds or substances through this pore according to their polar group affinities.

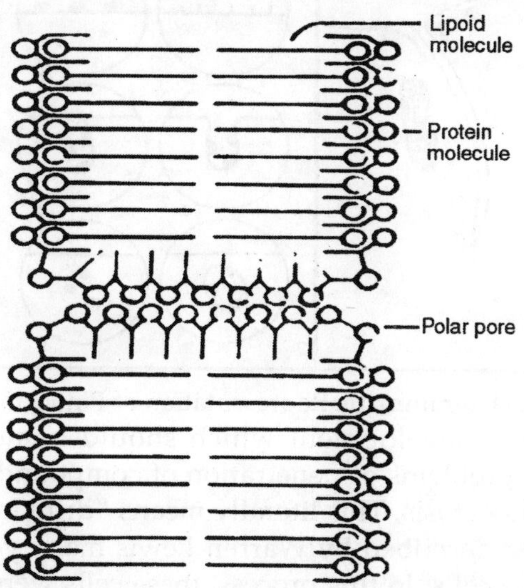

Fig. Diagram of Cell Membrane as Suggested

Danielli discusses the association with all membranes of a group of enzymes, which are called "permeases." These compounds he says "permit either facilitated diffusion or active transfer of 'signal' molecules across the membranes." It is possible that insulin may function as a permease for cells of certain organs. To quote Danielli again "Once permease becomes incorporated in the plasma membrane the situation is transformed: specific substrates can penetrate, more permease will become available by induced synthesis, induced enzyme will appear and a whole range of further induced enzymes may appear in response to the action of the induced enzymes of the inducing substrate.

Thus a transient infection with permease may potentially change drastically the state of differentiation of a cell—possibly even result in carcinogenesis." It is very likely that there will be considerable further developments in this subject of permeases and their relation to cell membrane permeability and these will be awaited with interest.

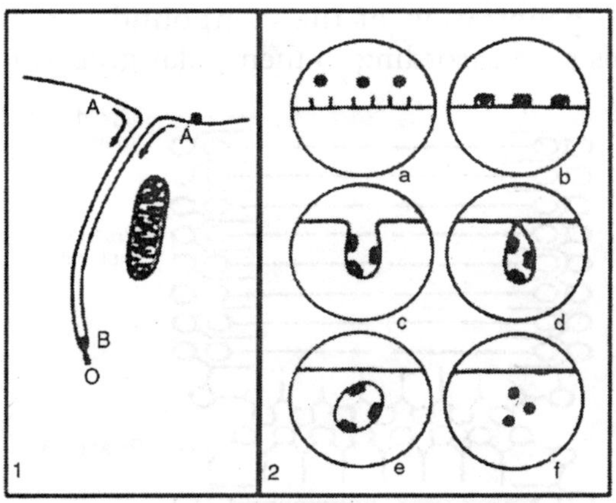

Fig. Diagrammatic Representation of Pinocytosis

A further development which should be noted when considering problems of penetration of compounds into cells is that of pinocytosis. This literally means "drinking by cells" and was first described by Warren Lewis many years ago in tissue culture cells. In this process, these cells were seen by a pseudopodial-like movement to engulf a droplet of culture fluid in which they were lying. The process was equivalent and similar to the phagocytosis of solid food by the *Amoeba,* only in the case of the tissue culture cell it is a droplet of fluid that is engulfed.

Now it seems that the cells of multicellular animals even in vivo can do the same sort of thing though there may be specialized parts of the cell membrane which carry out the process. A case in point is the kidney tubule cell. From the basal part of the cell, long internal extensions of the cell membrane pass a considerable distance up into the interior of the cytoplasm. They are present in some other cells too and are called "caveolae intracellulares."

The cell appears to be able to imbibe droplets of fluid by an engulfing movement of the membranes of these caveolae. It is possible too that the cell membranes of a variety of cells may be able to carry out pinocytosis even without the presence of these intracellular membranous extensions. Thus we see that, if a cell really "wants" to take in a few macromolecules that pass its membranes with difficulty, it can always take them by "the scruff of the neck" and pull them pinocytotically into its interior.

DIFFERENTIAL CENTRIFUGATION

One of the developments in technique which has done an enormous amount to advance our studies of the cell is that of differential centrifugation and an account of this technique should be given before we pass on to the study of the cell components.

The extraction of cell components from tissues dates back a very long time and studies on the chemical nature of the nuclei were carried out during the 19th century on nuclei obtained from pus cells by differential centrifugation. More recently, tissues have been homogenized and spun down at different speeds and the different portions of the cell separate out according to their density.

For example, if a homogenate is taken up in 0.8-M sucrose solution which tends to preserve the morphological character of most cell structures and it is then spun at about 1000 r.p.m., the cell debris and the nuclei come down. The nuclei can be differentially centrifuged again from the cell debris so that a pure nuclear fraction suitable for chemical analysis can be obtained. At a higher speed, the mitochondria and at still higher speeds, the particles of glycogen, one part of the microsomes and other parts which are really fragmented endoplasmic reticulum aggregate at the bottom of the centrifuge tube.

As a result of this technique, studies of the enzyme content of the nuclei and mitochondria have been made in great detail. This has caused some controversy because of the possibility that soluble enzymes and other substances might be leached from the nuclei during the homogenization procedures. Dilute acids, e.g., citric and acetic acids, have been used as homogenization fluids and possess certain advantages in that they make the tissue fragile and thus more easily broken. They also appear to harden the nuclear membrane and help to preserve the nuclei intact.

However, they are unsuitable for enzymatic studies and tissue homogenized in solutions of sucrose are usually used for this purpose. The latter method is claimed, because of the low ionic strength of the sucrose solution, to reduce greatly the loss of materials from the nuclei. It is said, in fact, that proteins and nucleic acids do not get through the membranes but low molecular weight substances do suffer some loss by this method.

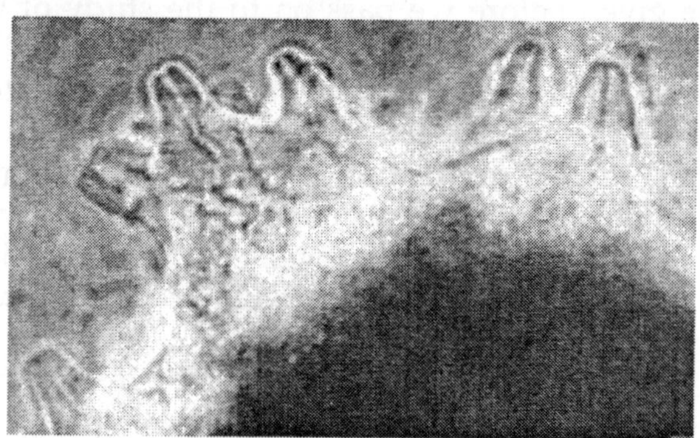

Fig. Phase Contrast Study of Pinocytosis in Amoeba

Allfrey and his colleagues at the Rockefeller Institute have tried to reduce any type of loss from the nucleus by freeze-drying the tissue and then grinding and floating it in nonaqueous media. Although in many respects this is advantageous, it appears difficult to be certain that small fragments of cytoplasm do not stick to the nuclei and contaminate the material. Extremely interesting results have been obtained by this particular technique.

Another check which was made on this differential centrifugation technique was to add various pure enzymes to the homogenate and then to calculate how much was adsorbed on the nuclei, mitochondria, microsomes and so on and how much was left in the supernatant. Although the amounts taken up with the supernatant varied with different tissues, generally speaking, the results were encouraging enough to indicate that the adsorption of enzymes from the normal supernatant was not really a significant cause of error in this technique.

Roodyn in a recently published review of the enzymic content of isolated nuclei has discussed the possibilities of contamination. He points out that we know that there is a very intimate connection between the endoplasmic reticulum and the nuclear membrane and that it is believed that the nuclear membrane is really a fold of the endoplasmic reticulum. Thus the nuclear fraction could easily be contaminated with this part of the cell.

Also there is a possibility that some of the sedimented material may contain whole cells which have escaped homogenization as well as nuclei and this would be another source of error. There is a possibility too that mitochondria may stick to the isolated nuclei and so also be contaminants. Although a certain amount of contamination is thus possible, it is only significant where a small percentage of total enzyme content of the cell is attributed to the nucleus, in this case it may be due to contamination, but where the nucleus contains an appreciable proportion of the total enzyme activity of the cell then the contamination becomes not very significant.

Chapter 3

The Cytoplasm

Within the cell membrane and separated from the outside (extracellular) world by it, is the cytoplasm. Does it perform any labour? What is its contribution to the life and welfare of the cell? First of all let us find out what we can about its nature.

Cytoplasm chemically is composed of proteins, lipoids (which include fatty, phospholipid and steroidal compounds), carbohydrates, mineral salts and, of course, a good deal of water, (40-80%).

The constituents of the protoplasm of animals and plants are generally similar, although the relative proportions of the various components varies a good deal in different organisms and probably even in the same organism under different physiological conditions. Originally, according to Wilson in his classic work *"The Cell"*, the similarity between the protoplasm of plants and animals was stressed and particularly the importance of protein as a structural element in both of them (this, of course, can still be accepted).

However, the chemical evidence has since shown that some of the other elements differ in animals and plants, for instance, the plant protoplasm has rather more carbohydrate in it and proteins and lipids are the main constituents of the animal protoplasm.

Protoplasm, in general, behaves and appears to be something in the nature of a colloidal system which is very complex and behaves almost always as if it were a viscous liquid. This is particularly well demonstrated by living protoplasm undergoing streaming movements so well demonstrated in plant cells and in animal cells, such as

Amoeba, when pseudopodia are being produced. Cells which are lying free tend to round up and become spherical when they are resting and if small fragments of protoplasm are chopped off from cells by a process described technically as "micrurgy," which simply means microsurgery, the little pieces so removed become spherical. Both these facts are in keeping with the conception that protoplasm is a viscous fluid.

Viscosity varies a great deal in different kinds of cells and it varies in the same cell from time to time according to the physiological state of the cell. Sometimes it may set into a semi-jellylike condition and this is particularly well shown by some types of slime molds (myxomycetes). When one of these is touched the whole organism suddenly sets into a gel formation and only after the passage of time does the organism gradually pass into a viscid fluid state again.

The ground substance of the protoplasm is known as hyaloplasm and contains a number of bodies and structures, vacuoles and so on and also a number of very tiny particles, some of them ranging into the ultramicroscopic level and which undergo active Brownian movement. According to Wilson "Flemming observed the dance of minute fat drops in living cartilage cells." Many of the theories of the nature of protoplasm developed in the latter part of the 19th century. At first there were not many theories about its structure because living cytoplasm is optically homogeneous or empty and between the years 1870 and 1890, when great interest was taken in examinations of fixed and stained tissues, a number of theories of the structure of protoplasm were developed based on its appearance after various types of fixation.

Probably one of the most wellknown of these is the fibrillar theory. This held that the protoplasm was made up fundamentally of delicate fibrils which were were either separate strands or formed a meshwork and this was situated within a homogeneous, optically empty, ground substance. Among the people whose names were associated with this theory were Leidig, Flemming, Carnoy and Heidenhain. These fibrils were thought to be of fundamental importance to the vital activities of the cell. However, later on, the fibrillar theory became subdivided into two subsidiary theories. One was the reticular theory and the other the filar theory.

The reticular theory was a modification of the filar theory in the sense that whereas the fibrils did not form networks in the filar theory, they did in the reticular theory. However, toward the end of the century a number of workers including Flemming, Bütschli and Fischer showed that coagulation artifacts could produce all the phenomena which had been described as fibrillar structures in the cytoplasm.

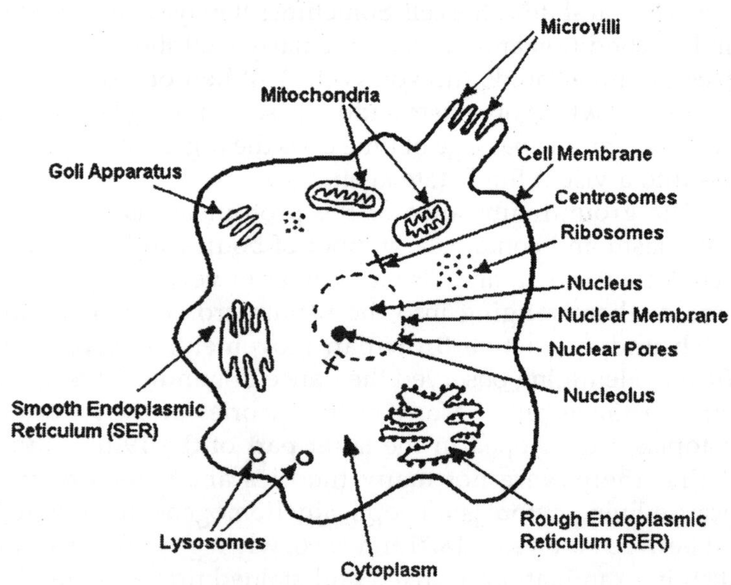

Fig. The Structure of Cytoplasm

A number of experiments had been carried out by various workers in which every conceivable type of fibrillar structure which had been described in the cell had been duplicated in non-living material made up, for instance, of white of egg or gelatine which had been coagulated by various fixatives. Thus by the end of the 19th century, the fibrillar theory and its offshoots were beginning to fall into disrepute.

Another theory of the structure of protoplasm was the alveolar or foam theory which was enunciated by Bütschli in a series of papers, the first of which he published in 1878. He believed that protoplasm was made up of a series of what he described as "alveolar spheres" which were suspended in a hyaloplasmic material. He regarded protoplasm as being equivalent to two viscid liquids, one of them forming the walls of the spheres scattered amongst the continuous substance (the hyaloplasm). There were also present in the cytoplasm a number of very small granules described as "microsomes," and we should note that the term "microsomes" as used by the early cytologists is quite different from the use of the term today. Nowadays "microsomes" have a fairly specific connotation since they are those bodies which are spun down in an ultracentrifuge from cell homogenates; the last of all the particles of the cell to be spun out. To what extent all modern microsomes represent something which really occurs in the living cell or not we shall discuss later on.

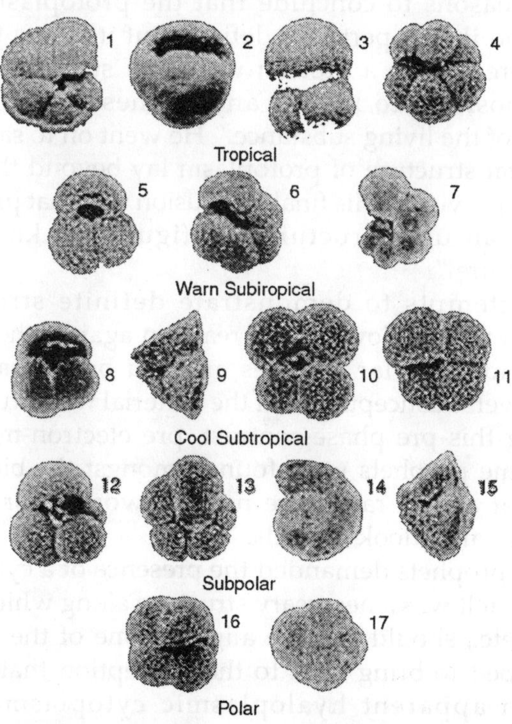

Fig. Granular Structure of Cytoplasm

Another theory which developed in the 19th century was the "granule" theory. This was introduced by Altmann in three papers in 1886, 1890 and 1894 and was subsequently developed by quite a number of other writers, mainly Benda and Meves. Altmann, using a special technique, demonstrated red staining fuchsinophil granules in all the cells he examined. In many cases they were a fairly uniform size and in some cells they were very closely crowded and appeared to occupy nearly all the cytoplasm. Altmann regarded these granules as elementary organisms and he called them "cytoblasts" or "bioblasts" and believed that they "lived" in a homogeneous ground substance again called the "hyaloplasm."

We know now, of course, that Altmann's granules were actually mitochondria. Wilson in 1928 summing up the theories of cytoplasmic structure said "we are driven by a hundred reasons to conclude that the protoplasm has an organization that is perfectly definite but it is one that finds visible expression in a protean variety of structures. We are not in a position to regard any of these as universally diagnostic of the living substance." He went on to say that the fundamental structure of protoplasm lay beyond their limits of microscopic vision. His final conclusion was that protoplasm possessed an ultrastructural configuration known as a "metastructure."

The attempts to demonstrate definite structure in protoplasm were followed by a reaction against the alveolar, fibrillar and granular theories and led back again to the extreme reverse conception that the material was structureless. Yet during this pre phase-contrast, pre electron-microscope period, some prophets were found amongst the biochemists who are, or at any rate were not noteworthy for having a morphological outlook on cells.

These prophets demanded the presence of a cytoskeleton which they felt was a necessary structure along which various enzymes, etc., should become aligned. One of the first steps which helped to bring back to the conception that structure existed in apparent hyaloplasmic cytoplasm was the development of the phase-contrast microscope in which a number of formed structures were seen to exist in the cytoplasm of the living cell. Immediately beneath the cell membrane, the protoplasm of the cell is differentiated to form an "ectoplasm" which appears to be in a partly gelled condition. It is probable that this ectoplasm plays an important part in the movements of the cell and it appears to be actively forming and reforming in the movement of protocells such as the Amoeba. It is also modified to form cilia. These are, in effect, ectoplasmic prolongations surrounded by a cell membrane. The ectoplasm is also a constituent part of the microvilli which occur on absorbent cells of the kidney tubules and intestine.

Phase-contrast microscopy demonstrated a good deal of structure in the apparent hyaloplasm and later this was supported by the results obtained with a further development of this sort of microscopy—the interference microscope in which objects of different density in the cell were coloured differently according to their density. More recently the electron microscope, using the excellent fixation produced by osmium tetroxide has demonstrated an incredible complexity in what was only a few years ago thought of as a relatively homogeneous hyaloplasm.

The cytoplasm is filled with mitochondria, microbodies of various sorts, double membranes, Golgi apparatus and various granules; in fact, there is relatively little of what we might describe as hyaloplasm. This hyaloplasm does, however, exist and when fixed and examined under the electron microscope it seems to be composed of an extremely fine network and this becomes coarser or finer according to the type of fixative used.

Now what is this network composed of? Wykoff has pointed out that a number of proteins which contain long, flamentous type molecules are capable of forming gels which have many of the physicochemical properties characteristic of protoplasm. He points out that gelatin is one of these and describes one of his experiments in which gelatin gels were first fixed with osmic acid and then examined under the electron microscope. He found that the gelatin was arranged in a network and that the pores of the net were small or large according to the concentration of the gelatin/gel or the fixative used. With a concentration of 2 to 4% in a gel, the net which results possesses a pore size which is very similar to that found in cells, but the pores get smaller if the gel is more concentrated. It is probable therefore that the network-like structure of the hyaloplasm which we see is really to some extent the effect of the fixative, although it is partly the expression of the nature and distribution of the filamentous molecules of protein which form the basis of the hyaloplasm.

These molecules are, in fact, probably arranged in a sort of network, a structure that Sir Rudolph Peters once described as the "cytoskeleton," and which he thought was the basis of cytoplasmic structure. This suggestion was made some years before the electron microscope demonstrated such a wealth of formed components within the hyaloplasm itself. Within the network of filamentous molecules (both protein and polysaccharide) which build up the hyaloplasm, a variety of molecules, not only free protein, polysaccharide and lipid but also ions, move about attaching and detaching themselves to the network in accordance with the electrical and other forces operating at that level.

We can, indeed, picture this hyaloplasmic network as a sort of skeleton for the rest of the cell. Now in this hyaloplasmic structure lie the rest of the cell elements which have to be separated from it to carry out the types of activity in which they specialize.

THE ENDOPLASMIC RETICULUM

A prominent feature of the cytoplasm of cells, particularly pancreatic cells, under the electron microscope, is a basophilic fibrillar structure which takes us back a little into the fibrillar theory of protoplasm; this structure is known either as the "endoplasmic reticulum" or "ergastoplasm" and reference will be made again to these names later on.

The structure and function of this material has been worked out and recognized in the last few years with the aid of the electron microscope, although it was originally discovered a long time ago with the light microscope. The discoverer was Garnier and the year 1897. The material discovered was a basophilic fibrillar material which could be seen in stained cells, particularly in the basal region of gland cells and it was called by Garnier "ergastoplasm."

```
          \            /
           C=O···H···N
          /            \
      RHC               CHR
          \            /
           N···H···O=C
          /            \
   ···O=C               N···H···
          \            /
           CHR    RHC
          /            \
   ···H···N             C=O···
          \            /
           C=O···H···N
          /            \
      RHC               CHR
          \            /
           N···H···O=C
          /            \
   ···O=C               N···H···
          \            /
           CHR    RHC
          /            \
   ···H···N             C=O···
          \            /
```

Fig. Hydrogen Bonds Between Polypeptide Chains

The ergastoplasm appears to be part of the system described as the endoplasmic reticulum. This latter structure was described by Porter and Thompson in a paper in 1947 on electron-microscope studies of a chick macrophage. The macrophage when very thinly spread out showed that the cytoplasm was everywhere permeated by a tenuous network which they described as a lacelike reticulum.

This they quite understandably called an endoplasmic reticulum. Later on, when fine sectioning of cells became possible in the 1950's, the endoplasmic reticulum seemed to occupy the position that was normally occupied by the "ergastoplasm," as Garnier had called it. In addition it also appeared to spread through the other parts of the cytoplasm of the cell and to be present, to a greater or less extent, in practically all the cells which have been examined. However, at this point we should try to solve the problem of terminology.

One of the characteristics of ergastoplasm is that it is basophilic, but not all the endoplasmic reticulum is basophilic although both basophilic and nonbasophilic regions in ultrathin sections of cells have the same double membrane structure.

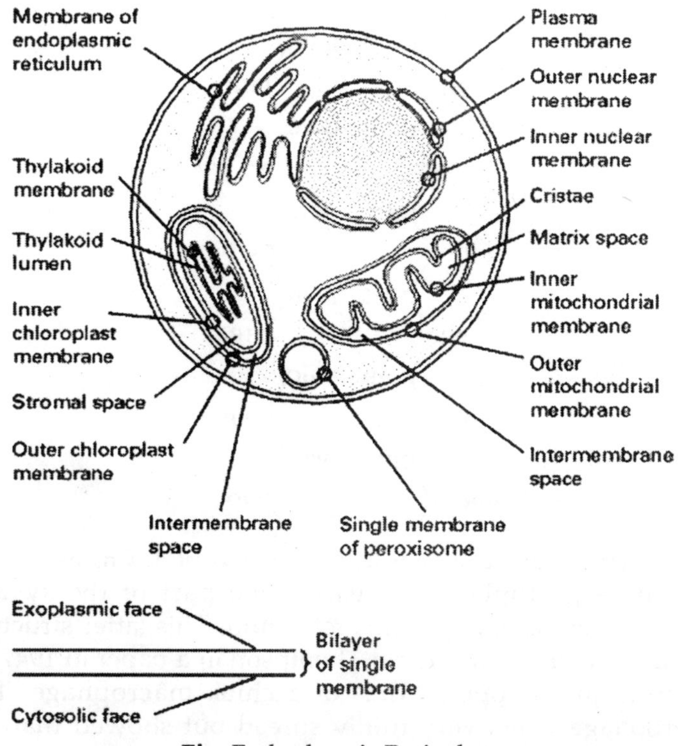

Fig. Endoplasmic Reticulum

The difference is that the basophilic portions have ribonucleoprotein particles attached to the membranes and the nonbasophilic parts do not. Therefore the ergastoplasm can be thought of as a specialized basophilic part of the endoplasmic reticulum. Sjóstrand tried to avoid this controversy by giving the names of Greek letters to various membranes in the cells. The membranes which form the endoplasmic reticulum he has described as the á-cytomembrane, the membranes of the mitochondria and of the cytoplasm receiving other Greek letters to designate them.

Sjóstrand's recommendation for nomenclature, however, seems to be a rather cumbersome method of describing these structures.

The use of the term "ergastoplasm" for the system of cytoplasmic membranes is largely supported by the French school recently headed by the late Charles Oberling, whereas in the United States the accent is more on endoplasmic reticulum with ergastoplasm as a specialized part of it. This is the nomenclature followed by the Rockefeller group headed by Porter and Palade.

Since the nature of the endoplasmic reticulum and, indeed, of many other cell structures is based on electronmicroscope studies, we might perhaps briefly consider the significance of electron-microscope pictures, in general and attempt to assess to what extent they represent a real structural condition in the living cell. The late Dr. Oberling has listed six points which suggest that the electron-microscope picture is a true picture of the living cell.

- The types of structures which are seen under the electron microscope have also been seen in living cells and they have been photographed or filmed by the phase-contrast microscope.
- Whenever the possibility has existed for comparison of the same cells, for instance, alive as viewed under the phase-contrast microscope or in fixed condition with the electron microscope, there has been perfect aggreement in the picture.
- The same methods of fixation and observation reveal all the structures side by side and present in all cells, even in those which are very different from both the phylogenetic and the functional points of view. The appearance of these structures depends to a large extent on the perfect preservation of cells by first class fixation.
- In order to attain excellent pictures the cells must be fixed in the living state. This proves the great sensitivity of the observed structures and the reliability of the procedures in detecting such

structural changes which take place immediately after vital functions have ceased.

- The techniques of homogenization, fractionation and ultracentrifugation have given the opportunity to isolate these structures and obtain them in a relatively pure condition in sufficient amounts to permit biochemical investigations, thus bringing closer the collaboration between the morphological and biochemical studies of the cell.
- These structures, as they appear under the electron microscope, do not always produce the same aspect, but they vary according to the evolutionary phases and also differ in pathological conditions from which the cell may have been suffering.

These points are very strong ones and possibly the strongest of all from the point of view of the very delicate structure which is demonstrated in electron microscopy is point 4. With regard to the first three, very few people have ever doubted that, for instance, mitochondria (as seen with the light microscope) existed in living cells. They have been seen in living as well as in fixed cells examined by the optical microscope and it was no surprise to find that they were also present in electron-microscope pictures. The point that has been really at issue is whether the extremely fine structures which the electron microscopists demonstrate do really exist in life.

The very fine, double membrane structures and so on, could possibly be produced as a result of the technique used, although in view of the widespread constancy of findings this is unlikely. Nevertheless we should still keep a healthy attitude of caution in accepting all these fine structural details and be prepared to alter our views should evidence accumulate that there are errors in this depiction of ultrafine structure. At the moment, however, it appears that the electron microscope is telling the truth so far as we can interpret it, but that it is not yet showing everything that is in the cell.

Garnier, the discoverer of ergastoplasm, brought out some very important points concerning its nature and these have been listed in an excellent article by Haguenau to which the reader is referred for further details of the ergastoplasm. Garnier, according to Haguenau, said that, in the basal region of all cells, without exception, there was a portion of the cytoplasm which seemed to have a fibrillar or rodlike structure, that these filaments or rods were not separate structures but were actually part of the cytoplasm and were in direct continuity with it. Haguenau points out that the electron microscope has provided an excellent confirmation of this. Garnier also said that these filaments were stained by basic dyes—those used were safranin gentian violet and toluidine blue—and that the intensity with which this basic staining occurred varied with the stage of secretion in gland cells.

Another point which he made was that the ergastoplasm was not a permanent structure and that its development was related directly to the state of activity of the cell. He said, for instance, that in gland cells the filaments appeared much more numerous when the cell, having already gone through a cycle of secretion and excretion, was preparing a new cycle. When the cell became loaded again with secretory granules, the filaments became less obvious and disappeared. His final point was that the filaments were closely related from a topographical point of view with the nucleus, that masses of the material formed laterally on the sides of the nucleus and that sometimes the latter was completely encircled by the ergastoplasm. Garnier also thought that nuclear sap or chromatic substance originating from the nucleolus was able to pass through the nuclear membrane and enter into association with the ergastoplasm. This was actually a very significant comment as will be seen later.

Following Garnier's work many other authors figured and described and categorized the ergastoplasm. Prenant in 1898 wrote a review on the ergastoplasm which he called the "protoplasme supérieur." He described the ergastoplasm as a very important zone of the cytoplasm which was capable of differentiating into specific structures and among these, according to Haguenau, were included the "Nebenkern," the

"Dotterkern" of the germ cell and the ergastoplasm of the gland cell and even the Nissl bodies of nerve cells.

These were rather interesting conclusions since they have now been supported by electron-microscope studies. Haguenau has pointed out that in the history of most discoveries there is a period in which it is first described, then a lot of other people describe it and finally there is a period when everybody believes it is an artifact due to fixation. She points out that at about the same time as the ergastoplasm was discovered Altmann discovered mitochondria, but comments made on the latter by Benda in 1898 and 1899 served to complicate the proper interpretation and acceptance of the ergastoplasm.

Haguenau divided the post-Garnier workers into three groups: the first group were represented by Morelle in 1927 who claimed that the ergastoplasm was nothing more than modified ground cytoplasm which took up basic stains, this being due not to any difference in structure but simply to a chemical difference. He claimed that this area of the cytoplasm was never really fibrillar. Then there was a second group who thought that the ergastoplasm was only mitochondria which were distorted in shape. Champy in 1911, for example, stated that mitochondria and ergastoplasm were one and the same and that the preparations which showed ergastoplasm were simply preparations with poorer fixation than those which showed mitochondria.

To some extent there is some justification for this point of view, since mitochondria and ergastoplasm are so closely related to each other topographically that it is natural enough for a mitochondrial stain to demonstrate concentrations of mitochondria in the same site as the ergastoplasm. The third group agreed with Garnier that mitochondria and ergastoplasm were quite different structures but these, according to Haguenau had to fight very hard to prove it. Prominent amongst these was Regaud.

The critical experiment which Regaud published in 1908 was one in which he showed that, if he had acetic acid in his fixative, there were no mitochondria in the preparation but the ergastoplasm appeared quite normal. On the other hand, with acetic acid absent mitochondria could be demonstrated very conveniently but the ergastoplasm did not take up its norm

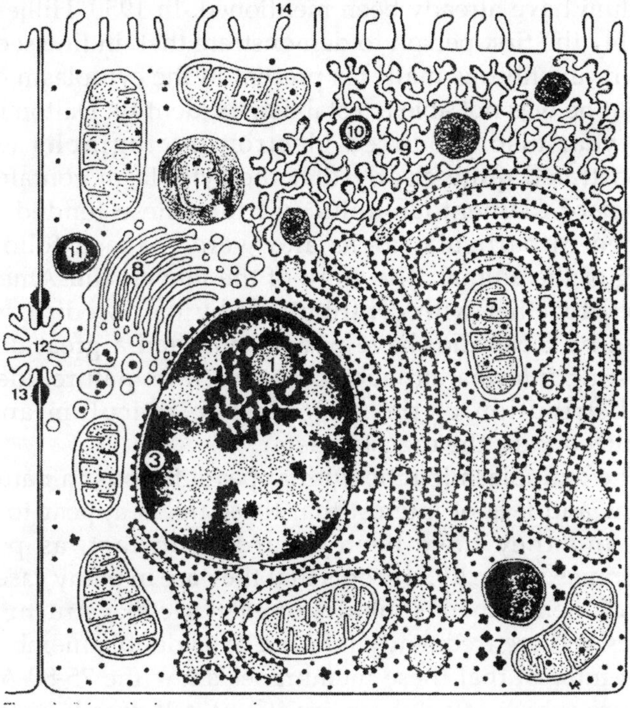

Fig. Endoplasmic Reticulum of Liver Cell 1. Nucleolus; 2. Chromatin; 3. Dense Chromatin; 4. Nuclear Pores; 5. Mitochondria; 6. Rough Endopla-smic Reticulum; 7. Ribosomes; 8. Golgi Apparatus; 9. Smooth Endoplasmic Reticulum; 10. Peroxisomes; 11. Lysosomes; 12. Bile Capillary; 13. Desmosomes; 14. Microvilli.

Haguenau quotes Regaud, "It is possible that the ergastoplasm consists of a protoplasmic support impregnated with chromatin or a closely related substance." Rees in 1940 using the polarizing microscope made a study of the ergastoplasm and came to the conclusion that there was a homogeneous filamentous structure present in the cytoplasm of living cells, but it was the electron microscope which finally confirmed and helped to delineate the structure of the ergastoplasm. Preliminary studies by Porter and colleagues leading to the designation of this material as endoplasmic reticulum have already been mentioned. In 1950, Hillier was probably the first person to demonstrate that, in fine sections of liver, fine fibrous matter is present in the cytoplasm of the cell, but he was not able to establish its identity. Dalton in the same year also produced electron micrographs which demonstrated quite clearly that the cytoplasm contained a number of filamentous units and that these tended to be grouped in particular areas and were reduced following fasting of the animal. The work of the French and American schools, together with contributions by the Swedish school headed by Sjóstrand, have now established a great deal of interesting information about the structure, nature, identity and distribution of the endoplasmic reticulum and its specialized part, the ergastoplasm.

The endoplasmic reticulum appears fibrous in nature in sections and it is of interest that these fibers appear to form pairs and they have been regarded, in fact, as paired membranes. Actually, it seems that they are in many cases the walls of membraneous tubules or even sac-like structures or vesicles and are frequently referred to as double membranes. It is of interest that these membranes show the 75-80-A unit structure which is so characteristic of cell membranes and which may be significant in the light of something we will say later on about them.

There is some discrepancy between the findings of electron microscopists regarding the ergastoplasm and the findings of the classical cytologists. The latter claim that the basophilic staining with which these fibrillar structures are associated appears to be localized at the basal end of the cell, whereas with the electron microscope these fibers or membranes extend all through the cell. This, however, can be explained in part by the fact that it is only when there are quite a number of membranes close together that they show up as a strongly basophilic area under the light microscope and in other parts of the cell the membranes are more widely spread out and are not organized in dense groups; therefore the characteristic staining reaction is spread out over a great area of the cytoplasm and becomes diluted.

Pancreatic Acinar Cell Model:

Ca^{2+} Regulated Signal Transduction Pathways

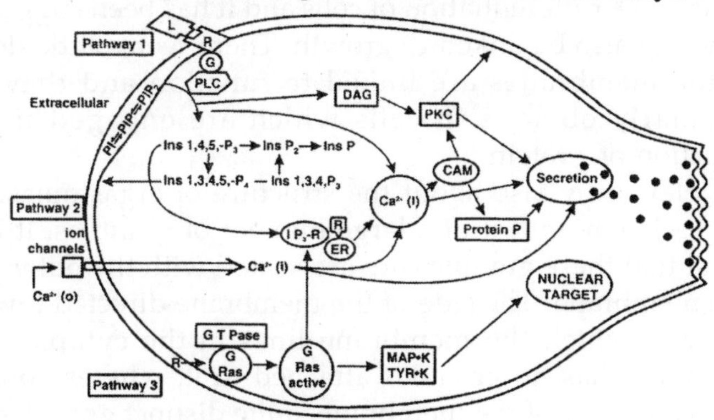

Fig. Portion Pancreatic Acinar Cell

Furthermore some of the membranes do not have ribonucleic acid granules attached to them (the component which produces the basophilic staining) and are called "smooth" endoplasmic reticulum. It is also evident that some endoplasmic reticulum membranes may show as basophilic structures but do not have any obvious RNA granules —so, perhaps, this material can be associated with the endoplasmic reticulum in a molecular as well as a particulate form.

In areas where the ergastoplasmic membranes are concentrated together so that the whole cytoplasm has a lamellar appearance, it has been suggested that the term "organized ergastoplasm" should be used. This term was suggested by Houdson and Ham in 1955. Another interesting organization of the ergastoplasm is the "Nebenkern." Nebenkern are structures which are basophilic in nature. Their significance and method of formation was unknown to older cytologists, but under the electron microscope it appears that they are concentric layers of ergastoplasmic membranes which in section look like an onion bulb or, as some authors put it, a finger-print.

It is possible that these structures may originate from mitochondria. The ergastoplasm may also be modified in nerve cells in the region where Nissl bodies are found, as demonstrated by Palay and Palade. There are considerable variations in amount of ergastoplasm which is thought to be related to the differentiation of cells and it has been suggested that it may also be linked to growth. There seems little doubt that the membranes are linked to function and they are particularly obvious in cells which are engaged in the production of protein.

Noteworthy also about the structure of ergastoplasm is the association with it of a large number of granules. It is of interest that these granules are associated with the outer part of the membrane. The side of the membrane directed toward the cavity which the membrane lines in the cytoplasm is smooth and has no granules attached to it, whereas on the cytoplasmic side of the membrane, quite distinct granules are found. They are basophilic in nature, range from approximately 120 to 150 A in diameter and have been the subject of considerable speculation.

Since they are basophilic in nature, it is obvious that they are, as mentioned earlier, the cause of the basophilic staining of the ergastoplasm recorded by the older cytologists. The studies by Palade and Siekewitz over the last five or six years have demonstrated the very interesting fact that these granules are composed of ribonucleic acid combined with protein. The way in which Palade and Siekewitz obtained this information is of interest. They treated homogenates of cells with deoxycholate, a surface-tension reducing agent which detached the granules from the membranes. Then they were able, by differential centrifugation, to isolate a number of them, carry out chemical analyses and so establish that they were largely composed of ribonucleic acid. However, one should not think that all the ribonucleic acid in the cytoplasm is associated with the endoplasmic reticulum. It is present in other parts of the cell as well, but there is some evidence that about 25% of the cytoplasmic ribonucleic acid is, in fact, associated with the endoplasmic reticulum.

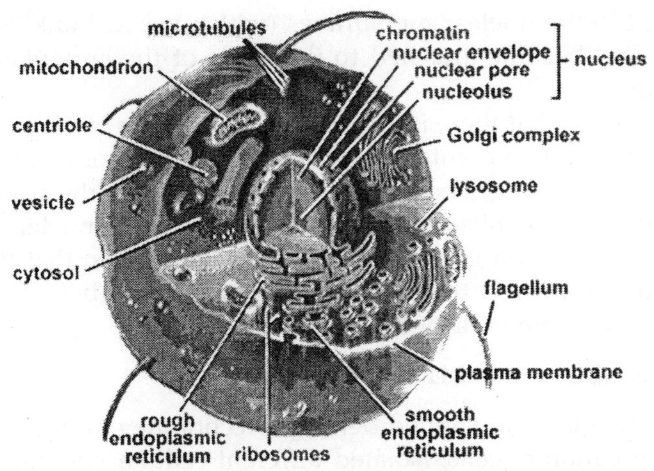

Fig. Hypothetic Cell

One of the interesting observations about these particles, demonstrating perhaps not so much division of labour in cells as co-operation of labour between cell constituents, is that in the nucleolus, ribonucleic acid particles are found which are the same size (150 A in diameter as those associated with the endoplasmic reticulum.

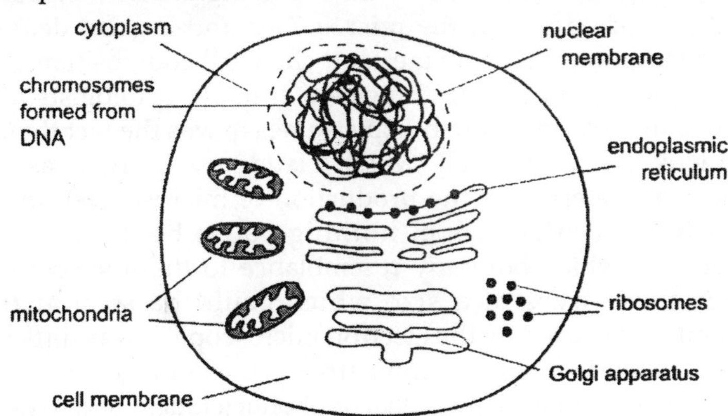

Fig. Principal Organelles in a Cell

(It is of interest, also, that the deoxyribonucleic acid particles which are found in the chromatin of the nucleus are also about 150 A in diameter.) It seems possible that the RNA particles could pass out from the nucleolus through pores present in the nuclear membranes (which we will talk about later) and become attached to the walls of the endoplasmic reticulum.

The fact that there is a coincidence in size is of very great interest from this point of view. A different type of migration has also been suggested by Gay who showed with the electron microscope that blebs forming at the surface of nuclei may become converted into endoplasmic reticulum and that RNA granules appeared to become attached to them before they actually became detached from the nucleus.

THE NATURE OF MICROSOMES

When Claude applied Bensley's technique of differential centrifugation to cells, isolated different cellular components and subjected them to chemical analysis, he described a series of minute bodies which are submicroscopic in size, i.e., below the limit of the resolution by the optical microscope. He described these bodies as microsomes. They are the smallest bodies in cellular homogenates and come down after everything has been centrifuged off, i.e., at the highest speed of centrifugation, leaving only the supernatant. Many chemical studies were done on the microsomes and a good deal of information on their enzymes and chemical composition was obtained. However, one thing which made cytologists and particularly electron microscopists uneasy was the fact that, if one looked at a liver cell (which Claude used largely as the source of material for the production of microsomes) under the electron microscope, nothing could be seen in the cytoplasm which bore any resemblance to the microsomes. Since these were of a size which could be seen at the magnification used by the electron microscope, it was difficult to understand where they came from in the homogenates. The microsomes contain a good deal of ribonucleic acid and so does the ergastoplasm.

Finally, by isolating the pellet of microsomes centrifuged out from the homogenized cells and by examining ultrathin sections of them under the electron microscope, various workers showed that the microsomes were small vesicles with ribonucleic acid granules attached to them. It thus seemed quite obvious that the grinding-up of the cell during homogenization broke up the endoplasmic reticulum and the bits formed themselves into little vesicles which floated in the fluid and were finally centrifuged down. This work was first done by Slatterbach in 1953 and subsequently by Palade and his colleagues.

Relation of Endoplasmic Reticulum Membranes to Cell Membrane

The E.R. (endoplasmic reticulum) membranes have a structure which is similar in structure and dimensions to the cell membrane. It has been suggested that the E.R. membranes might represent complex infoldings of the cell membrane and that the cavities between the folds of the membranes may be in direct continuity with the exterior of the cell. We know that some infoldings occur, they have been mentioned before and are known as the caveolae intracellulares. These structures have a space of constant size between the two folds of cell membrane which form their walls—this space is 200 A in width, thus the total width of the system including the two membranes of 80 A thickness each is about 360 A. Now, in some parts of the cell the membranes and spaces of the endoplasmic reticulum approach this dimension, but there is tremendous variation in the space (known as the "cisternal space") between the membranes, in parts it may be greatly dilated.

This suggests that the endoplasmic reticulum is not necessarily part of the same system to which the caveolae intracellulares belong. No certain connection between the E.R. membrane systems and the cell membrane has been established although the possibility exists and Porter and other authors have recognized that the connection may not be permanent but that it may be spasmodic. A distinct possibility occurs also that the endoplasmic reticulum developed originally from infoldings of the cell membrane, that it then increased in complexity and finally lost its connection with its point or points of origin, but temporary direct connection between the cisternae of the endoplasmic reticulum and the exterior of the cell can occur.

It is thus possible that the cavity between the membranes in the E.R. represents a path by means of which the contents of the membranes or sacs (cisternae) can be excreted to the exterior. In other words the secretory products of the cell pass from the cytoplasm through the walls possibly in molecular form, collect into droplets in the interior of the membranes and pass along them. It is, of course, possible that secretion products form in the cisternae. In 1957, Hendler and colleagues found that, in the gland cells which produce albumen in the hen oviduct, the cavities or spaces between the membranes were dilated and contained a precipitate which possessed the staining and other characters of the secreted material which is found in the lumen of the gland. There is other evidence to indicate too that, possibly, these spaces between the membranes represent a pathway for secretory material.

It is of interest, as shown by Brandes that, in the ventral lobe of the prostate gland in mice and rats, the spaces between the pairs of membranes are so enormously dilated that the membranes from opposite pairs come into close apposition with each other; pairs of membranes still exist but they are the halves of opposite pairs which have come together. The space between the original pairs of membranes is packed with an amorphous material. The ribonucleic acid granules in the such pairs of membranes are, of course, on the inside instead of the outside as they were when the original halves of the pairs were together. It is possible that this enormous accumulation of material in these endoplasmic sacs really represent accumulation of secretion.

It is of interest that, if the animals are castrated, the pairs of membranes with the granules on the inside separate from each other, the amount of secretory material appears to decrease and the original halves of the membrane pairs come close together again so that the ribonucleic acid granules can now be seen on the outside. Brandes has also demonstrated that, by injection or implantation of male sex hormone, the original state of the cell could be produced again. This is in a sense an " Alice Through the Looking-Glass" type of cell in which the normal arrangement of cytoplasm and endoplasmic reticulum spaces is reversed. When this cell is in its normal condition, the spaces between the membranes contain, in fact, the strands of cytoplasm and by far the main body of the cell is occupied by what are presumably the secretion products.

It has probably not been made clear up to date that all these endoplasmic sacs are probably continuous with each other and thus form a convoluted system of spaces coursing through the cytoplasm of the cell. It appears, also, that folds of the endoplasmic reticulum constitute the nuclear membrane (this will be discussed later). Since this membrane has been shown to contain pores, it is quite possible that material could pass from the nucleus into the cisternae and along these channels to the exterior of the cell without having to traverse the cytoplasm or the cell membrane at all.

At this point we should enlarge a little on the subject of the caveolae intracellulares. These were originally described by Yamada as consisting of a number of infoldings of the surface membrane of some cells. He found them in endoneurial, endothelial, pulmonary epithelial and muscle cells. Although it is possible that these caveolae could, in fact, represent the regions where the E.R. channels communicate with the exterior, the existing evidence is against this. If this were so then there is a continuous, if extremely tortuous, pathway which stems from the vicinity of the nuclear membrane extending to the exterior of the cell and which could be described as a possible circulatory system for the cell.

At the moment it seems most likely that the caveolae represent only blind pockets which extend for variable distances into the cells and there is no evidence that they are actually continuous with the ergastoplasm. Palade has suggested that these invaginations in endothelial cells, at any rate, represent stages in a process of what is described as "micropinocytosis" which has already been discussed. It has been suggested that when these little vesicles containing water pass into the cell, the cell membrane then dissolves away and the water is liberated into the cytoplasm.

This conception of the cell membrane "pinching off" and a little vesicle passing into the cytoplasm is attractive, but one of the difficulties of accepting it is the mechanism of the formation of the vesicle. We know that the cell membrane can invaginate in this way (it has been seen to do so) but whether, when the membranes on the two sides of the vesicle bend around and come in contact with each other, they are, in fact, capable of coalescing and nipping off a vesicle is another matter. It has been pointed out that, if the membrane of a cell was a purely lipid substance, there would be no difficulty at all in such a concept, but, since the outside and probably the inside parts of the lipid membrane of a cell are covered with protein, this would make such a coalescence rather difficult to conceive from a physicochemical point of view.

On the other hand, an invagination of the membrane may draw off a little droplet of water and the membrane may then burst internally and squirt the drop of water into the cell and then there would be no difficulty in the burst sides of the membrane joining up in the usual way. This seems to the present author to be a much more likely way for pinocytosis to take place than for a vesicle to be actually cut off.

The Golgi apparatus has also been shown to be composed of double membrane structures without, however, the associated ribonucleoprotein granules of the endoplasmic reticulum. Although we will be dealing with the Golgi apparatus later on, we might mention here the concept which has been put forward by some authors that the Golgi apparatus represents a part of the system of the endoplasmic reticulum and that it is in communication with it. We do not know whether the whole of the endoplasmic reticulum is in continuous communication with itself, however, if it is and if the Golgi apparatus is part of it, then it is possible to assume that everything passing along the cisternae of the reticulum will have to filter through the Golgi region. It should be stressed that there is no proof that this is so, that many cytologists believe it is not so, but that such a possibility exists.

One problem which should be discussed at this point is the origin of the endoplasmic reticulum. Where in fact does it come from? There are many controversial theories on this subject. It has been suggested that it might originate as an infolding or infoldings of the cell membrane and this theory was put forward by Palade.

The Nebenkern because of its high concentration of E.R. membranes might be considered its a possible region in the cytoplasm where this material is being formed. The Nebenkern is made up of a dense aggregation of concentric rings and one could, perhaps, consider the possibility of the membranes peeling off from such a germ centre. The same sort of appearance has also been noted in association with the nuclear membrane and has led to the suggestion that the endoplasmic reticulum is formed in this region and that the layers are, in fact, peeling off the nuclear membrane. However it may quite easily be interpreted the opposite way, i.e., the close apposition of the existing ergastoplasm around the nuclear membrane may be a part of the complex canalicular system which permits an almost direct passage through the nuclear membrane direct into the cisternae of the reticulum. Again, there are a number of authors who think the endoplasmic reticulum is associated with mitochondria.

There is no doubt that many pictures of mitochondria, closely surrounded by concentric layers of endoplasmic reticulum have been obtained and we, ourselves, have found this particularly well demonstrated around the mitochondria of the liver cells of scorbutic guinea pigs. In some cases, Sheridan in this department has found a tremendous concentration of many layers of reticulum around the mitochondria as though, in fact, the reticulum is being formed on the surface of the mitochondrial membrane and is being split off. Rouiller and his colleagues have demonstrated that, in animals which have been poisoned, the endoplasmic reticulum, destroyed or badly damaged by the treatment, always reappears in association with mitochondria.

The membranes of the reticulum may not undergo direct physical formation in the sense that the membranes are produced on the surface of the mitochondrial membrane and split off; it may be that the mitochondria supply the energy necessary for the production of these membranes. Furthermore the close relationship between the endoplasmic reticulum and the mitochondria may simply be physiological, that is they are co-operating in some metabolic process such as the synthesis of protein. However, the origin of the endoplasmic reticulum is a problem which is far from solved and we must await further evidence before drawing a conclusion.

Chapter 4

The Mitochondria

Mitochondria have been known for many years. They were first discovered and described by Altmann in 1886 and were put more or less definitely on the cytological map by Benda in 1903. They can be easily demonstrated with suitable dyes. Altmann's aniline fuchsine-picric acid technique shows them up very well. Regaud's iron-hematoxylin method is also very good. The mitochondria can even be seen in the living cell by staining them intravitally with Janus green and, of course, they show up extremely well with the phase-contrast and interference microscopes. There is not much difficulty, then, in establishing their existence and nature. Their dimensions vary, of course, but they range in most cells from about 0.5 to 2ì or longer.

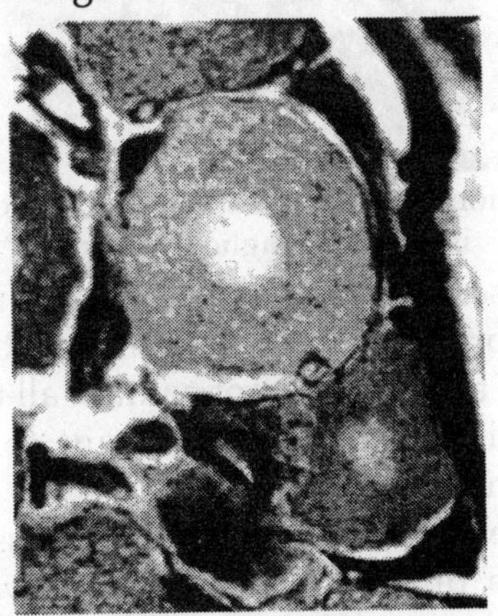

The filamentous mitochondria, for example, which are present in connective tissue cells may be much longer than this. Extremely small and even submicroscopic mitochondria may also exist. Extremely smallsized mitochondria are, in fact, known and Rhodin has described structures ranging from 0.1 to 0.5ì which are presumably mitochondrial in nature. Green has pointed out that the various enzymes and proteins which mitochondria contain can be accommodated in a very much smaller particle than any known mitochondria and still perform all the functions required of them.

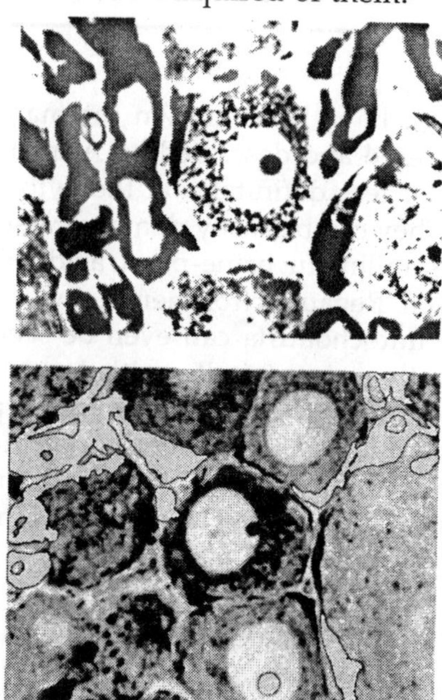

Fig. Relation between Nucleolus and Mitochondria in Spinal Ganglion Neurons.

Some of these much smaller mitochondria may not have all the details of organization of larger mitochondria but there is no reason why they should not contain all the enzymes and other compounds which are important for mitochondrial structure and function.

It is of interest to note that, in living tissue culture cells, the mitochondria undergo continuous movement. This has been noted by quite a number of workers and the movements have been described frequently. These movements may be fitful individual movements by the mitochondrion itself, i.e., bending, wriggling and writhing, or may consist of movement of the whole mitochondrion through the cytoplasm of the cell. The cause of the bending and twisting movements of the mitochondria is not exactly known. At one time it was thought to result from the contracting of the polypeptide chains which are part of the membrane of the mitochondria. It may be due to the fact that the mitochondria were engaged in active ion pumping or being twisted by cytoplasmic movements. In the second type of movement in which there is a transport of the whole mitochondrion from one part of the cell to another, the movement is possibly related to streaming movements of the cytoplasm or to electrical forces.

The mitochondria have been described as making journeys from the cell membrane to the nuclear membrane and back again almost as if they were discharging either an electric charge or even perhaps some compound on the membrane that they touch. There are also records that, when the mitochondria come in contact with the nuclear membrane the nucleolus sometimes moves across the cell and comes in contact with the nuclear membrane at the same point at which the mitochondrion is touching it and this suggests that there may be some exchange of compounds or substances during this period. This has been demonstrated in spinal ganglion cells by Dr. Tewari in the author's laboratory. Filamentous mitochondria have frequently been observed to break up into batonettes and these have been observed to break up into granules. The reverse process has also been found to occur, granules have been seen to join up into batonettes and these into longer filamentous mitochondria. What the significance of this breaking up is we do not know, but a little later we will refer to this process again in the light of what will be said about the metabolic activities of mitochondria.

When mitochondria were studied in ultrathin sections of tissue, it was found by Palade in 1953, Sjöstrand, 1953 and by Rhodin in 1954, that they had an interesting internal structure. These workers found first of all that there was a single membrane round the outside, that inside this membrane was another membrane so that the two made a double membrane and, after a little controversy, it was agreed that the inner of these two membranes was extended to form a number of bars or plates which projected into the interior of the mitochondria, in some cases touching or almost touching the other side. These plates or bars or tubes had double structure again, since they were a reflection of the inner membrane of the mitochondria.

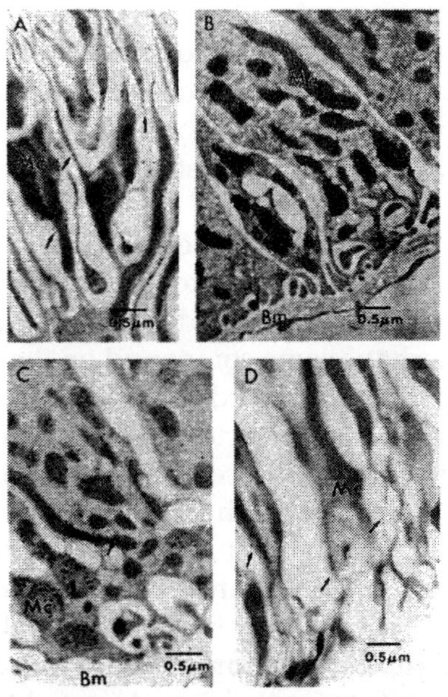

Fig. Elongated Mitochondria in Kidney

They were described by Palade as the "cristae mitochondriales". Changes in these cristae appear to be related to function, for instance Palade has drawn attention to the fact that the amount of cytochrome in the mitochondrial fraction of a homogenate is directly related to the number of cristae present in the mitochondria.

When metabolic activity is high, for example, in rapidly contracting skeletal and heart muscle, the mitochondria have many densely packed cristae. In smooth muscle, however, where the activity is greatly reduced, the cristae present in the mitochondria are relatively sparse. There are claims that, e.g., in mouse tumor cells and in paramecium, cristae may be everted into the cytoplasm. Presumably in this case the outer membrane is folded into cristae.

It is of interest that the double membranes which surround the mitochondria show the same 80 A unit structure which is characteristic of the cell membrane and Robertson has suggested that it is possible that the mitochondria may have originated by an invagination of the membrane which was nipped off to form the mitochondrion.

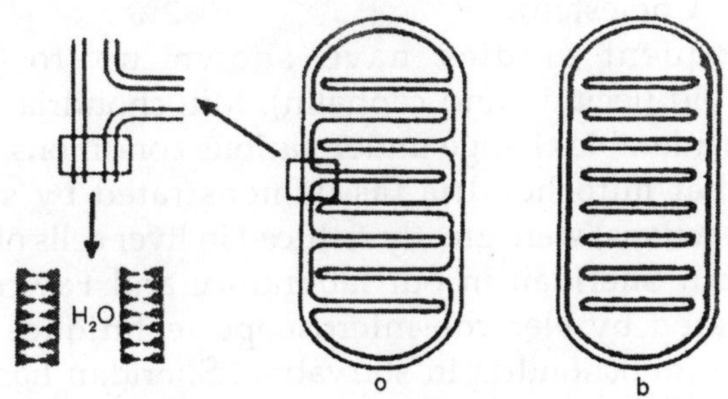

Fig. Structure of Mitochondria

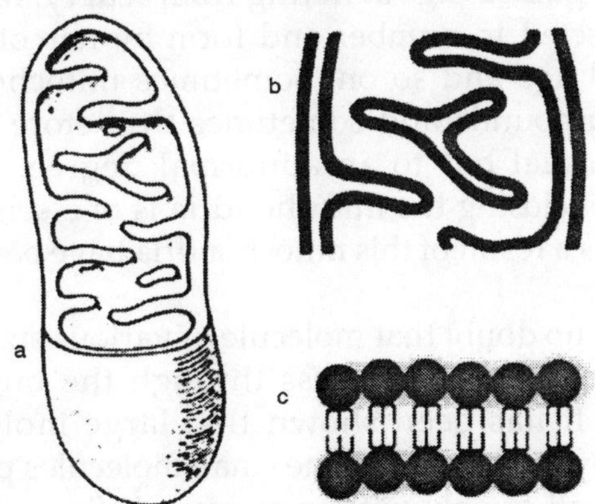

Fig. Three Dimensional View of Structure of Mitochondria which Appears as a Bag filled with Fluid.

This is an interesting suggestion but it is not readily acceptable to most cytologists and is, in fact, considered unlikely by many of them. The membranes of the mitochondria appear to consist of two protein layers separated by a double layer of lipids. This is again characteristic of cell membrane structure. Bensley in 1934 was the first one to isolate mitochondria by differential centrifugation and he was able to demonstrate that mitochondria contained a considerable amount of protein, lipid and fat. Bensley's chemical composition of mitochondria is

Proteins and unknowns	65%
Glycerides	29%
Lecithin and cephalin	4%
Cholesterol	2%

Subsequent studies have shown up to 30% of phospholipid (lecithin and cephalin). Mitochondria undergo a very considerable change under various conditions. Bensley claimed that mitochondria (as demonstrated by standard staining reactions) were greatly reduced in liver cells of starved animals, but Sheridan in our laboratory and Fawcett have demonstrated by electron-microscope techniques a great increase of mitochondria in starvation. Sheridan has further demonstrated a still further increase of mitochondria in the liver cells of guinea pigs suffering from scurvy. Mitochondria are also affected in number and form by narcotics, enzyme poisons, old age and so on. Sometimes mitochondria store unusual compounds and sometimes they store compounds which are usual but to an abnormal degree. The double membrane enclosing the mitochondria is of a semipermeable nature and as a result of this mitochondria have been described as osmometers.

There is no doubt that molecule of various sizes pass with different degrees of readiness through the mitochondrial membrane. It has been shown that large molecules pass through fairly easily while some small molecules pass through either not at all or only with very great difficulty. It has even been said by Emmelot and Bos that this permeability, particularly of liver mitochondrial membranes, is affected by thyroxine. Also there are claims by some authors that mitochondrial membranes have pores. Mitochondria may accumulate water and swell. They may do this in starvation and in various pathological conditions. When this occurs they produce the condition which the pathologists describe as cloudy swelling. Cloudy swelling has been known to occur sometimes in neoplasms and sometimes after poisoning with various toxins.

The mechanics of cloudy swelling are that the mitochondria become very large and lose their cristae, the matrix stains less strongly and sometimes the outer membranes of the mitochondria may disappear and the mitochondria may fuse and so form structures known as chondriospheres. Under some pathological conditions mitochondria may become disrupted, undergo lysis and portions may be ejected from the cell. Mitochondria are capable of abnormal storage. Accumulation of ferritin by mitochondria was demonstrated by Kuff and Dalton in 1957 and they may also accumulate iron pigments or bile pigments. Brachet has described ferritin particles in young erythroblasts where he says they penetrate the mitochondria and then break into smaller iron-containing particles. When the mitochondria eventually burst, the granules become scattered throughout the cytoplasm and take part in the formation of hemoglobin. Silver granules and keratinous materials have been also seen inside mitochondria. They have been found to contain melanin and Graffi in a series of papers from 1939 to 1941 claims that they also contain or store carcinogenetic hydrocarbons. Sometimes this storage affects the function of the mitochondria and sometimes the accumulation of material is due to malfunction of the organelle. They may also store neutral fats or lipids. In plants mitochondria have been recorded as storing starch.

As a result of storage mitochondria undergo physical changes. The membrane may become single and the cristae decrease in size and may be lost altogether. If the stored substance is eliminated, the mitochondria resume their normal appearance. There is a vast literature on the role of mitochondria in the production of almost any kind of product—protein, yolk, fat, glycogen and so on—but this is not the place to review this work.

There have been many scornful remarks in the literature about the supposed synthetic activity of mitochondria, but the study of the enzymic equipment of mitochondria shows that they have the equipment to effect a wide range of synthetic activities. It has been mentioned that the isolation of mitochondria by the differential centrifugation of homogenized cells was first carried out in 1935 by Bensley and by Hoerr. It was not until 10 or 15 years later that this technique became used extensively and this time by biochemists who set to work to find out the metabolic composition of mitochondria, i.e., the nature and content of enzymes concerned with metabolic processes.

THE CHEMICAL NATURE OF MITOCHONDRIA

In Bensley's work it was suspected that mitochondria contained a good deal of lipid material. One finding which suggested this was that mitochondria were stained by a method which is similar to that used for staining *Mycobacterium tuberculosis* and *Mycobacterium leprae*. These bacteria are stained by treating them for some minutes with a hot phenolic solution of basic fuchsine and once stained in this way they resist the de-staining effects of acid alcohol. This is the origin of their designation as acid-fast bacteria and their staining idiosyncrasies are believed to be due to a waxy or lipoidal coat. Mitochondria stained by hot acid fuchsine resist the decoloring action of picric acid for longer periods than most other cell constituents and this suggests that the mitochondria might also have a membrane which contains material of a lipoidal nature or at any rate which is of similar composition to that of the bacteria mentioned.

Mitochondria are difficult to demonstrate with conventional methods of staining if the fixative has contained acetic acid, alcohol, ether, chloroform, acetone, or other fat solvents. Also mitochondria have been found to stain with osmic acid and, incidentally, to resist de-staining by extraction with turpentine—a property of certain types of lipoprotein complexes. Baker has found that his "acid-haematin" test for lipids applied to a variety of tissues, nearly always gave a positive, reaction with the mitochondria. Thus there had accumulated for some years direct and indirect evidence that mitochondria contained appreciable amounts of lipid. Bensley's experiments, which, incidentally, were carried out some years before Baker's staining studies, showed that mitochondria which had been produced by homogenization and differential centrifugation contained appreciable amounts of lipid (30%).

METABOLIC SUBSTANCES FOUND IN MITOCHONDRIA

Perhaps we should give a brief summary of what was suspected about the metabolic significance of these organelles prior to the revolutionary studies which were made with isolated mitochondria beginning during the middle of the 1940's. One of the characteristic staining reactions of these organelles is that they give a green or green-blue stain with Janus green B which is diethylsafranine azodimethylaniline and it has been claimed by Cowdry that this reaction is primarily due to the diethylsafranine monocarboxylic acid component since this compound gives a very good and specific reaction with these structures. It was observed many years ago by T. B. Roberston that if one drop of a saturated solution of safranine was added to a solution of trypsin, a coloured precipitate was formed and subsequently it was demonstrated that this coloured precipitate had proteolytic activity.

Then Marston demonstrated that other azo dyestuffs would react in a similar way, particularly neutral red which is a dimethyldiaminotoluazine hydrochloride. Marston suggested that these results indicated that the reaction of mitochondria with Janus green might signify that they contained proteolytic enzymes. Now, the concept of the staining reaction with Janus green has been pretty well proved by Lazarow and Cooperstein to indicate that mitochondria play an important part in cellular oxidations and that the production of a pink colour from the Janus green by the mitochondria is due to the DPN specific dehydrogenases. The fact that Janus green B does not stain the mitochondria permanently green but that the green colour gradually becomes reduced to the pink and then to the colourless form has been known for many years. Joyet-Lavergne demonstrated more than twenty years ago that mitochondria contain an oxidase system which oxidizes cobaltous to cobaltic salts and that these latter stain the mitochondria green. It was noted by Gatenby that in the snail, *Lymnaea,* the mitochondria are coloured yellow in the natural state presumably by a carotenoid pigment; also extracted lipomitochondria frequently have a yellow appearance which too is probably due to carotene. It is of interest in this connection that it has been demonstrated by the present author and by Joyet-Lavergne that mitochondria give a blue reaction with antimony trichloride in chloroform solution, an acknowledged reaction for vitamin.

A. Carotene is also a provitamin A, so these histochemical results indicate the presence of vitamin A in mitochondria. Criticism of Joyet-Lavergne's results was made by Gomori because the former author had used alcohol as a fixative. The present author has, however, always applied antimony trichloride in chloroform solution direct to fresh unfixed tissues and this suggests that this result is a true one, in any case biochemical tests have now confirmed that mitochondria (prepared by homogenization and differential centrifugation of the cells) contain 27-32% of their weight of lipids and that 100 mg. of this lipid contains approximately 249-910 U.S.P. units of vitamin A. JoyetLavergne has suggested that mitochondria contain a redox system in which vitamin A plays a part.

A number of authors, Leblond, the present author and Giroud and his co-workers have demonstrated that mitochondria of some organs react with acetic acid-silver nitrate solution (which has been demonstrated as being a specific reagent for vitamin C) to give a positive reaction and this indicates that vitamin C may be present in them. One has to accept the intracellular localization of vitamin C with a certain amount of discretion in view of the very destructive effect of this reagent on the cell cytoplasm. Electron-microscope studies in the present author's laboratory have demonstrated that this reagent has a most drastic effect on the ultrastructure of the cell and we cannot be sure that the localization of vitamin C in the mitochondria is, in fact, a real thing.

It is of interest, however, that Chayen has found that the mitochondria of plant cells give this reaction very intensely and very specifically. It may be that mitochondria in plant cells and some animal cells do, in fact, contain vitamin C, but we need further studies before this can be confirmed. Other vitamins appear to be present in mitochondria, members of the vitamin B complex have been found to be present, in some cases, in the form of coenzymes. The actual vitamins of the B complex recorded are vitamin B1 (thiamine), riboflavin, nicotinic acid (niacin), pantothenic acid and pyridoxine.

However, although these vitamins are present in mitochondria, they are not present in any greater concentration than in the other parts of the cell so that they are not exclusively contained in these organelles. These studies were made on cell homogenates and, of course, it is possible that the presence of the vitamins in the other fractions of the homogenate may be due to the fact that they have been leached out of the mitochondria by the saline solution which is used in the homogenization process in these particular experiments and further work would have to be done before we could be certain about this.

The present author, Joyet-Lavergne and Giroud have demonstrated that mitochondria contain glutithione or protein-bound SH and this is further evidence that mitochondria can play an important part in oxidation-reduction mechanisms in the cell. It is of interest that mitochondria have been found in large quantities in the phloem cells of plants which are concerned with transport and may be concerned with this process. This suggestion is made in view of Conway's views that redox systems can play an important part in ion pumping. Another substance which was demonstrated histochemically in mitochondria by the present author, using the Schultz reaction, was cholesterol. This was particularly evident in the mitochondria of cells of the adrenal cortex. Similar but less intense reaction was shown by the mitochondria of the liver and it has been demonstrated that isolated liver mitochondria contain about 2% cholesterol and it is possible they contain more in the adrenal cortex.

Bensley has recorded the presence of a red pigment in the mitochondria and also in the submicroscopic particles of the liver cell. He believed that this pigment is derived from the oxidation of unsaturated fats and possibly the phospholipids of the mitochondria and this led him to suggest that in the liver cell the mitochondria in particular are, possibly, the seat of highly active oxidative processes which involve the metabolism of fats. The possible relation of mitochondria to oxidation-reduction processes was indicated by the publications of Ludford. He demonstrated that, if methylene blue was added to tissue cultures, the mitochondria of the cells stained a brilliant blue colour but this could be inhibited by potassium cyanide. If the cells were exposed to a bright light, the blue colour was rapidly bleached.

It is of interest that the mitochondria, although they have been demonstrated to contain a good deal of fat and lipid, do not give a positive reaction with Sudan III. They contain protein (quite a high proportion of it), but it is of interest that the earlier workers, using Millon's reagent, produced a negative reaction for protein. Subsequently Bensley and Gersh using a Millon's reagent of a different formula were able to obtain positive results from the mitochondria of many tissues and they found that they were particularly well stained in frozen, dried sections and particularly in those of *Amblystoma* liver. The same authors demonstrated that, in undenatured, frozen, dried sections of *Amblystoma* liver, the mitochondria were destroyed if the sections were exposed to artificial gastric juice and artificial pancreatic juice.

Despite all this chemical information most of which was in existence by the beginning of the 1940's, there was no certainty as to the function of mitochondria and it was not until 1946 and 1947 that the late George Hogeboom and his colleagues at Bethesda completely revolutionized our ideas of their function by demonstrating that the major proportion of the succinic dehydrogenase and an appreciable proportion of the cytochrome oxidase activity of the liver cell were present in the mitochondria. This at once suggested that these organelles were the major sites of aerobic respiration in the cell. It is of interest that as long ago as 1915, Dr. Kingsbury had suggested that mitochondria were concerned with the respiration of the cell. His reasons for this were largely due to his observations that anesthetics such as ether and chloroform, which depressed cellular respiration and the respiration of the animal in general, also broke up mitochondria in the cell.

The work of Hogeboom and his colleagues was carried out on homogenates of liver which had been produced by grinding up liver with saline. This gave poor preservation of the form of the mitochondria and subsequently 0.8 *M* sucrose was used—this preserved the nature and form of the mitochondria very well. In the beginning there was some doubt as to whether the material being assayed was in fact mitochondria or not, but when sucrose was used this doubt gradually disappeared. Eventually electron-microscope studies of the structure of the granules isolated by homogenization and differential centrifugation demonstrated beyond doubt that they contained the same structures as the mitochondria of normal cells.

Fig. Localization of Respiratory Chain Enzymes (black spheres) in the Mitochondrial Membrane

Another possible source of error, however, began to haunt

the biochemists and was frequently verbalized by cytologists and this was that the mitochondria did not really contain the oxidative enzymes, mentioned in the foregoing, but that they were being absorbed or adsorbed by them from the homogenate. This possibility was negated at least in part by adding more enzyme to the homogenate and demonstrating that it could be recovered almost 100% from the supernatant and so was not taken up by the mitochondria. Thus the localization of the enzymes in the original cell was probably in the mitochondria.

Subsequently, Kennedy and Lehninger demonstrated that in addition to containing cytochrome oxidase and succinic dehydrogenase, mitochondria also catalyze the condensation of pyruvic acid with oxaloacetic acid, the formation of succinic acid from á-ketoglutaric acid and the formation of malate from succinate. These three reactions represent three steps of fundamental importance in the Krebs tricarboxylic acid cycle of which we will have more details shortly. Mitochondria also contain coenzyme I of which nicotinamide is an important constituent and cytochrome reductase which is a flavoprotein. These two enzymes are links between the Krebs tricarboxylic acid cycle and the cytochrome system and it, therefore, appears that mitochondria probably contain the whole enzymic equipment necessary for aerobic respiration of the cell. In fact, it was originally demonstrated that a centrifugate of cells which included only nuclei and mitochondria were capable of carrying through the whole of the oxidation of glycogen to CO_2 and water; subsequently it was demonstrated that isolated mitochondria on their own could do this. Transaminase activity was subsequently demonstrated in mitochondria although in only the same concentration as in the rest of the cytoplasm.

Transaminase is concerned with protein synthesis and it is of interest that pyruvic, oxaloacetic and á-ketoglutaric acids are the corresponding á-keto acids of the amino acids, alanine, aspartic and glutamic and need only to be transaminated to produce them. Since the former compounds are themselves formed as intermediates in the course of the Krebs cycle, this indicates a mechanism whereby mitochondria might synthesize fresh protein material and so increase in size themselves or synthesize protein for other parts of the cell. This possibility is further extended by the fact that RNA is known to be present in varying amounts in mitochondria, although the exact amount is uncertain and there is a possibility of contamination of mitochondria with RNA from the rest of the homogenate when making such estimates. However, there is no doubt that mitochondria do have the equipment for synthesising protein although to what extent they do this *in vivo* is not known.

All the fatty acid oxidase activity in the cell is also in the mitochondria and so is 80% of the octanoxidase activity. Since these discoveries, an enormous amount of work has been done on the mitochondrial enzymes and this work was summarized by Hogeboom (unfortunately now deceased), Kuff and Schneider from the *National Institutes of Health* and has been published in Volume 6 of *The International Review of Cytology*, 1957. It may be of interest to record here other enzymes and compounds which they listed as being present in mitochondria.

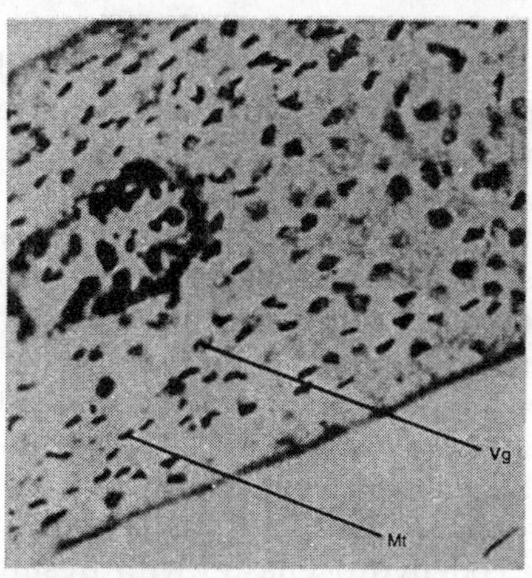

Fig. Mitochondria in the Protozoan Opalina.

It is of interest too, that a considerable amount of ribonuclease and deoxyribonuclease are present in mitochondria—this seems unusual since only a very small amount of DNA and a relatively small amount of RNA are present in the mitochondria compared with other parts of the cell. However, it is possible that the bodies which contain RNAase and DNAase are not true mitochondria but are lysosomes; we will discuss these shortly. In the meantime, it is of interest to note that it was eventually demonstrated by fragmenting mitochondria and separating the membranes from the contents that cytochrome oxidase and succinic dehydrogenase activity were located primarily in the membranes.

Subsequently it has been shown that the ability of the mitochondria to carry out the Krebs cycle and to combine this with oxidative phosphorylation is dependent on the presence of intact double membranes in the mitochondria. Now, it has been demonstrated and mentioned earlier that in the structure of mitochondria there is an outer membrane and an inner membrane to the mitochondrion, so that it is surrounded by a double membrane. Also, the inner membrane is folded inward to give elongated cristae and these cristae, because they consist of a fold of the inner membrane, are themselves double membranes. There appears to be little difference biochemically between the double membrane structures of the cristae and the double membrane structure of the outer wall of the mitochondria.

An important factor in the process of oxidative phosphorylation is the transport of electrons. Dr. David E. Green has described the process of electron transport as carried out fundamentally with the aid of an electron transport particle—this is a submitochondria particle.

The latter can be isolated in a form so that it can no longer carry out the full citric acid cycle, but it can still carry out the process of oxidative phosphorylation which it performs by the oxidation of succinate and DPNH (reduced diphosphopyridine nucleotide, a codehydrogenaste) with freshly prepared by homogenization and differential centrifugation and are then exposed to alternate freezing and thawing, their form undergoes considerable change. First of all, the external membrane ruptures and some of the cristae break off and eventually the mitochondrion is completely disintegrated; however, fragments of double membrane structure are left which, according to D. E. Green, may be derived either from the cristae or from the outer envelope of the original mitochondrion. These double membrane fragments represent collections of Dr. Green's electron transport particles.

If other methods are used for preparing these particles and the particles so produced do not show a double membrane structure, then oxidation of succinate and DPNH coupled with phosphorylation does not occur. Apparently the double membrane structure is essential for this process to take place. In the structures which do not have the double membrane structure, the capacity for electron transport is still present but oxidation phosphorylation cannot take place.

About 35% of the dry weight of the electron transport particle of Green is composed of lipid and the lipid is concentrated in packets of lipoprotein. These are interspersed between two molecules of cytochrome or between a flavoprotein and cytochrome. In other words, all the oxidation reduction members of the election transfer chain are connected to one another by lipoproteins. These lipoproteins, according to Green, act not only as structural devices, holding various enzymes together, but they also conatin a compound which can shuttle electrons back and forth so that then can pass from cytochrome through to flavoprotein and reverse. This compound, which Green and his colleagues have described, is a new coenzyme called coenzyme Q.

It is a completely water insoluble benzoqtuinone derivative, which is connected in the lipoproteins of the electron transport particles and is related to vitamin K. Coenzyme Q can undergo oxidation and reduction in a similar fashion to cytochromes; for instance, the quinone part of the molecule can be reduced to become a hydroquinone and the hydroquinone can be reoxidized. It is pretty certain that this coenzyme plays an important part in the electron transfer chain of the particle. The significance of the electron transport in these mitochondrial membranes may seem obscure but, as

Green points out, one should think of this problem of electron transport in terms of its coupling—the fact that the process can be coupled to the esterification of inorganic phosphate. In this process a monophosphoric ester of the hydroquinone or, what Green described as a semiquinone form of coenzyme Q would be formed.

This ester would then react with adenosine diphosphate by a process of transphosphorylation and ATP would be forme and simultaneously the semiquinone or hydroquinone would be oxidized to the quinone form in coenzyme Q by the ferric form of cytochrome c. The production of ATP is, in fact, the reason for the existence of such a system. This explanation of the way in which the oxidative processes of the mitochondria actually are used to produce ATP by oxidative phosphorylation is an extremely fundamental discovery and Green and his co-workers are to be complimented very much on the extraordinary patience and care which they have put in over a large number of years to work out this very complex problem in mitochondrial physiology.

In view of the studies by Green and his colleagues which have just been described, it is of interest to consider some of the studies of Martias. This author has for some time put forward the view that there may exist in mitochondria two routes for the transport of hydrogen ions or electrons, respectively, between the pyridine nucleotides (DPN and TPN) and cytochrome c. He thought that only one of these would be linked with enzymes which brought about phosphorylation. The route which is best known leads from DPN to cytochrome c reductase and then to cytochrome c. Martias believes that this particular route is not one which is concerned with the formation of highenergy phosphates.

On the other hand, he proposed a second alternative route in which vitamin K (phylloquinone) is an important factor. Until recently this concept of an electron transport system involving vitamin K in mitochondria has not proved to be particularly popular, Martias and Strouff have finally isolated a phylloquinone reductase that is a flavoprotein having a number of similarities to cytochrome c reductase. There is, however, a difference between them. Phylloquinone reductase and cytochrome c reductase can both react in the reduced state with phylloquinone, but with cytochrome c only cytochrome c reductase will have any effect. Vitamin K reductase has an extremely high activity even in relatively low concentrations.

If we consider that there are two routes in mitochondria which can be used for electron transport, then the problem to be solved is what decides whether these electrons and hydrogen atoms are directed into either of these particular pathways. Under normal circumstances they seem to be directed exclusively along the pathway which is coupled with oxidative phosphorylation and which leads via vitamin K reductase to vitamin K, cytochrome b and so through to cytochrome c.

This pathway via vitamin K seems to be sensitive to any alteration of the internal structure of the mitochondria, possibly because some of its constituents, particularly the vitamin K, are associated with the double membranes of the cristae. If their combination is disturbed, the electrons take the emergency route which is via cytochrome c reductase to cytochrome c. It is of interest in connection with this theory that dicoumarol is capable of decoupling oxidative phosphorylation from respiration and, since dicoumarol is known to be a vitamin K antagonist, its action is possibly explicable by the fact that it blocks the pathway between vitamin K and its reductase.

It is also of interest that thyroxine is a hormone which causes uncoupling of respiration from oxidative phosphorylation. Since it is known that, if thyroxine is added to mitochondria, it causes a swelling and change in the internal structure, this would be perhaps further evidence that the internal structure of the mitochondria is the one which is particularly concerned with the flow of electrons along the route leading to oxidative phosphorylation. It may well be that hormones which affect respiration do so by affecting the structural nature of the mitochondria, apart from their effects on the endoplasmic reticulum which have already been mentioned. The precise relation between Green's coenzyme

Q and Martias' vitamin K phylloquinone reductase is not absolutely clear, but it is very likely they are identical or closely related systems.

LYSOSOMES

De Duve in his studies on ultracentrifugation of cytoplasmic particles has demonstrated that some particles appear to contain acid phosphatase, cathepsin and ß-glucuronidase and ribonuclease, DNAase and cholinesterase. It is of interest that these compounds appear to be present in separate particles—they require higher centrifugal forces for sedimentation than do the cytochrome-oxidase bearing mitochondria. De Duve has pointed out that it is of special interest that such hydrolytic enzymes are in special particulate bodies differing from others of the cytoplasm. He is not clear as to how this should be interpreted but has pointed out that, if hydrolytic enzymes are free to act within the living cell, as, for instance, they can do in homogenates, they would interfere with the efficiency of the synthesis and may even affect the structural integrity of the cell; he then suggests that segregation of hydrolytic enzymes in this way is one method by means of which this activity is either kept in check or localized in specific parts of the cell.

De Duve once described these bodies as "suicide bags" and suggested that on the death of the cell these enzymes are released completely into the cytoplasm and play a part in the process of autolysis. The present author has suggested that these enzymes may be less effectively contained in the lysosomes in senescence and that this leakage may be partly responsible for the cellular process of aging. It is of interest that lysosomes have been demonstrated *in situ* with the electron microscope and stained to demonstrate their acid phosphatase activity. No typical mitochondrial structure can be seen in these bodies and they appear to be of a different nature from mitochondria.

THE METABOLISM OF CARBOHYDRATES

The role of mitochondria in oxidative phosphorylation has

already been mentioned and their role in the metabolism of carbohydrates through the presence in their substance of enzymes concerned with the Krebs cycle and the cytochrome system has been indicated. We should now consider the problem of carbohydrate metabolism and the role that mitochondria and other parts of the cytoplasm play in it.

Carbohydrate metabolism is extremely important for cell synthesis and is the main source of energy for cell activities. Before attempting to localize the various activities of carbohydrate metabolism in the actual parts of the cell, we should consider briefly what the metabolism of carbohydrates involves. There are two types of metabolism, anaerobic and aerobic. The anaerobic route is demonstrated very well by muscle and most of the information on this type of metabolism of carbohydrates has been obtained by studies of this tissue.

The result of anaerobic metabolism is the production of lactic acid and the liberation of a good deal of CO_2. However, although we think in terms of anaerobic metabolism for muscle, we have to realise that muscle itself has a first class blood supply which appears to be increased by various physiological mechanisms when muscle is forced to do work and that muscle in the process of contraction uses a rather surprisingly large amount of oxygen.

Lactic acid has been shown to accumulate in muscle extracts and in isolated muscles kept under anaerobic conditions. If we consider the accumulation of lactic acid in an animal in vivo, we find that after moderate work the accumulation of lactic acid goes up slightly but remains at a pretty steady level. However, if strenuous work is done, then the amount of lactic acid goes up extremely steeply and slowly comes back to normal. The reason for this is that under normal circumstances muscle can obtain oxygen fast enough to reoxidize the lactic acid as rapidly as it is formed and only a small amount of lactic acid accumulates. However, it is possible for muscle to do more work than it can supply oxygen for and it can do this by oxidizing carbohydrates anaerobically and so accumulating lactic acid. Eventually this lactic acid has to be converted with the aid of oxygen but this can take place over a longer period. Some of the lactic acid is converted into glycogen in the liver and the rest is oxidized.

The fact that it is possible to accumulate lactic acid and slowly oxidize this after the work is finished provides a mechanism by means of which an "oxygen debt" can be produced. Under aerobic circumstances it is not lactic acid which is formed in the metabolism of carbohydrates but pyruvate. However, this does not accumulate and it is oxidized almost as rapidly as it is formed—as we shall see in a minute there is a very complicated system for oxidizing this pyruvate. The only time when pyruvic acid does accumulate in the tissues is in the absence of vitamin B_1 or thiamine, a fact which was demonstrated years ago in Oxford by R. A. Peters. Under anaerobic conditions pyruvic acid undergoes the process known as oxidative decarboxylation with the aid of the cocarboxylase (thiamine pyrophosphate). It yields acetate, carbon dioxide and lactate. It has been stressed that this reaction is fundamentally of an oxidative nature and leads to the term oxidative decarboxylation which is in important process both in carbohydrate and protein metabolism.

We have mentioned the production of pyruvate but have not yet considered the process by means of which this compound is produced—this process is known as glycolysis and it represents the first stage in the metabolism of carbohydrates. Glycolysis is, in effect, the reverse of photosynthesis. In photosynthesis the energy of sunlight is used to combine CO_2 and water into carbohydrates; in the process of glycolysis, the glucose which is formed from carbohydrates such as glycogen and other polysaccharides is converted into CO_2 and water with the liberation of energy. We can express this as $C_6H_{12}O_6 + 6\,O_2 = H_2O + 6\,CO_2 +$ energy. This oxidation of glucose to give CO_2, water and energy is not a single step but involves a very large number of steps of considerable complexity. During the various steps, small packets of energy are released at a rate at which the cell can use them, whereas if there was a sudden explosive release of energy by the oxidation of glucose the cell would probably not be able to use this relatively large amount of energy in a coordinated way and a good deal of it would probably be wasted. The first stages of glycolysis involve the phosphorylation of glucose and this is done with the aid of ATP as follows (the enzyme concerned in this process is placed above the arrow):

$$\text{glucose} + \text{ATP} \xrightarrow{\text{hexokinase}} \text{glucose-6-phosphate} + \text{ADP}$$

Hexokinase is not just one enzyme, there are in fact a number of hexokinases that catalyze phosphorylation of hexoses. The phosphorylation of glucose results in the transferring of a high-energy phosphate from the ATP to glucose and so to form a phosphate ester which is poor in energy—this type of reaction is called an exergonic reaction and is essentially irreversible. It is of interest that one of the properties of glucose-6-phosphate which differs from glucose is the fact that the phosphate ester has difficulty in penetrating cell membranes whereas glucose itself, apparently, crosses without any difficulty and it has been suggested that this hexokinase reaction is one way in which glucose can be locked in a cell. It is also an essential prerequisite for the resynthesis of glycogen.

Many things can happen to glucose apart from phosphorylation and ultimate conversion into CO_2 and water via the glycolytic and Krebs cycle system. Amongst these are its dehydrogenation by a glucose dehydrogenase to form gluconic acid. This is done by a diphosphopyridine nucleotide (DPN) linked (codehydrogenase) reaction. In mammals, it is believed that this is not a usual pathway for glucose to follow. If glucose can be locked into position in a cell by being converted into a phosphate, it is obvious that there must be in existence a mechanism which can release it again since it needs, for instance, to be fed from the liver periodically into the blood stream to keep the blood glucose level at a relatively constant figure. This is carried out by a specific enzyme, glucose-6-phosphatase,

$$\text{glucose-6-phosphate} + H_2O \xrightarrow{\text{G-6-Pase}} \text{glucose} + PO_4.$$

This glucose-6-phosphatase is probably present in all

tissues which release glucose from cells, but it does not appear to occur in skeletal muscle.

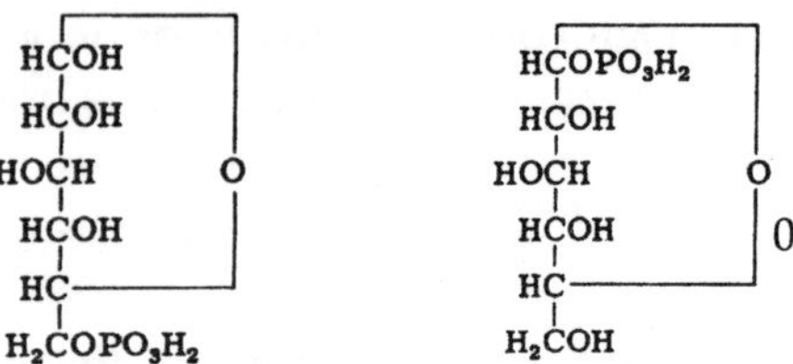

Fig. Glucose-6-phosphate α-Glucose-1-phosphate

Glucose-6-phosphate may be converted into glucose-1phosphate, in other words the phosphate is simply shifted around from the 6-position to the 1-position on the glucose molecule.

This change of position of the phosphate group is catalyzed by an enzyme known as phosphoglucomutase and the reaction changing the phosphate group from one part of the molecule to the other is an easily reversible reaction. The formation of glucose-1-phosphate from glucose-6-phosphate is a stage in the synthesis of glycogen, for instance, if glucose is accumulating in a cell it is converted to glucose-6-phosphate and then to glucose-1-phosphate.

The molecule of glucose-1-phosphate then polymerizes into glycogen. The reverse process can occur—glucose-6-phosphate can be obtained from glycogen by first producing glucose-1phosphate and then by producing glucose-6-phosphate from the glucose-1-phosphate. We have, however, diverted a little from the direct line of our story of the glycolytic cycle.

After the formation of glucose-6-phosphate from glucose, the next stage in the glycolytic cycle is the conversion of glucose-6phosphate into fructose-6-phosphate. This reaction is readily reversible and is catalyzed by an enzyme called phosphohexose isomerase.

Fructose-6-phosphate may also be formed directly from fructose and it is known that an enzyme, fructokinase, which is present in brain and muscle can produce this effect. The next step is the further phosphorylation of fructose-6-phosphate; another phosphate group is added in the 1-position to give fructose 1-6-diphosphate. The enzyme responsible for this is phosphofructokinase and the reaction is carried out with the aid of ATP:

fructose-6-phosphate + ATP $\xrightarrow{\text{PFkinase}}$ fructose-1-6-diP + ADP

This reaction is also exergonic: Fructose-1-6-diphosphate is also known as hexose diphosphate and in the next stage this is broken down by what is described as the "aldolase" reaction into 3 phosphorylated compounds—triosephosphates. The first of these is ketose triosephosphate, the second is dihydroxyacetone phosphate, the third is phosphogly ceraldehyde. The triosephosphate can be converted into phosphogly ceric acid; dihydroxyacetone phosphate can be converted into phosphoglyceraldehyde or alternatively it can be reduced to á-glycerophosphate with the aid of DPNH and á-glycerophosphate dehydrogenase. This reaction is potentially important for the synthesis of lipids since from the á-glycerophosphate, phosphatidic acid can be synthesized and phosphatidic acid can be the starting point for the synthesis of lecithin, cephalin and also of fats. Phosphoglyceric acid with the aid of the enzyme enolase becomes converted into phosphoenol-pyruvic acid.

Pyruvic acid penetrates cell membranes very well and can thus leave the cell and in theory can be distributed in any cell in the body. Conversely all the steps which we have mentioned can be reversed and pyruvate can be converted back into glucose-6-phosphate. When pyruvic acid is reduced it gives lactic acid. Lactic acid itself can be converted back to pyruvic acid. Liver cells are capable of reversing the whole glycolytic series of reactions and can produce glucose and glycogen again from lactic acid. Muscle can reverse lactic acid to glucose-6-phosphate and glucose-1-phosphate and glycogen but cannot produce nonphosphorylated glucose.

In these first stages of carbohydrate metabolism, a certain amount of energy is liberated but this is not much more than about one-tenth of the total amount of energy which is produced by the production of CO_2 and water from glucose. The glycolytic part of the cycle is the least energy producing part. In addition to its conversion into lactic acid or its oxidation, pyruvic acid can be converted to alanine, an amino acid, by the process of transamination. Thus here one can see a link between carbohydrate and protein metabolism.

If there is ample oxygen, the pyruvic acid produced by this first stage of carbohydrate metabolism can be oxidized. This is a complex process in which in the first stage the pyruvic acid is converted by the process of oxidative decarboxylation into acetyl coenzyme A and CO_2. Thiamine pyrophosphate (cocarboxylase) is an essential enzyme for this process.

Acetyl coenzyme A is a 2-carbon substance and it condenses with oxaloacetic acid which is a 4-carbon dicarboxylic acid to yield a 6-carbon tricarboxylic acid, namely, citric acid. The enzyme catalyzing this is called the "condensing enzyme." Thus starts a series of changes which ultimately lead to the formation of CO_2 and water. The production of citric acid is followed by the loss and recapture of water and it becomes converted to cis-aconitic acid (with the aid of aconitase, glutathione and ferrous iron) which on further hydration is converted into isocitric acid. This compound then loses hydrogen and thereby becomes oxidized to oxalosuccinic acid (this reaction is catalyzed by isocitric dehydrogenase) and decarboxylation turns it into á-ketoglutaric acid (the enzyme responsible is oxalosuccinic decarboxylase and oxidized manganese). á-Ketoglutaric acid is decarboxylated and then oxidized by the loss of two hydrogen atoms to succinic acid (the enzyme concerned is á-ketoglutaric dehydrogenase).

Succinic acid is also oxidized by the loss of H_2 (with the help of succinic dehydrogenase) to fumaric acid. The latter by the addition of the elements of water becomes converted by fumarase into malic acid and the malic acid by dehydrogenation (enzyme malic dehydrogenase) is converted in oxaloacetic acid and there we are back at the beginning of the cycle again. The oxaloacetic acid is ready to combine with another molecule of acetyl coenzyme A to produce citric acid once more. During this process 3 molecules of CO_2 are given off and 5 molecules of H_2.

It can be seen that quite a complex series of reactions take place in what has been called the "tricarboxylic" or "citric acid cycle," or the "Krebs cycle." All three terms are applicable. The final combination of the hydrogen atoms liberated by the Krebs cycle with oxygen is brought about by the cytochrome system. It is of interest that cytochrome is a protein which contains a form of heme, an iron containing pigment which is also present in hemoglobin. It is probably more widely distributed than any other type of heme protein since it occurs in the cells of all organisms that use oxygen, irrespective of whether they are animals or plants or whether in the case of animals they are vertebrates or invertebrates, or protozoa.

In the cytochrome system a variety of compounds and enzymes are involved. Cytochrome oxidase is a widely distributed enzyme since its occurrence is comparable in distribution to cytochrome. It has not yet been obtained pure and it has been found to be bound to the insoluble material when cells are homogenized and spun down. This is not surprising when we appreciate that cytochrome oxidase is associated with the mitochondria. There is also another compound involved called flavoadenine dinucleotide (FAD) for short) which is associated with various proteins to form a variety of different enzymes. A flavine containing enzyme was first isolated and called a yellow ferment as long ago as 1932 by Warburg and Christian.

Flavoproteins are usually associated with metals of various sorts and molybdenum, copper and iron are three which have been found to be necessary for their adequate functioning. It is not quite clear where the metal atoms are situated on the molecule but they appear to be essential for the action of the enzymes with which they are associated. There are two types of riboflavin containing enzymes, one type is an electron acceptor from reduced DPN or TPN and it can transfer these either to oxygen or to the cytochromes, whereas the other type of flavoprotein accepts electrons directly from metabolites. Among the important flavoproteins is cytochronie reductase which comes in two types—one for reduced DPN (DPNH) and the other for reduced TPN (TPNH). The oxidation of these two compounds (DPNH and TPNH) in the cell can thus be carried out by the cytochrome system. Now we are in a position to describe the next stages which take place in the oxidation of glucose. The 5 pairs of hydrogen atoms which are passed down to the cytochrome system from the Krebs cycle are combined with the cytochrome and result in its reduction. The enzyme concerned is cytochrome reductase, the flavoprotein already mentioned. Cytochrome is then oxidized with the aid of cytochrome oxidase, which removes the hydrogen atoms. They become combined with atmospheric oxygen to produce water and thus the long journey of oxidation of carbohydrates is done, 5 molecules of water and 3 molecules of CO_2 being produced from each molecule of pyruvate.

An analysis of oxygen uptakes of various tissues in the body gives an indication of the degree to which their cells are metabolizing. As a matter of interest, it might be noted that of the tissues examined, retina and kidney had the highest metabolism and liver was next, then the rate decreased progressively from adrenal, lung, bone marrow, diaphragm, heart, lymph nodes, skeletal muscle, skin to eye lens which had the lowest level of oxygen consumption. The important point to remember is that the complex of reactions described above is localized to a great extent in the mitochondria.

One of the important functions of this oxidative cycle just described is its relationship to phosphorylation. We have described earlier the work of Green who showed that oxidative phosphorylation could take place without the whole Krebs cycle occurring, but in intact mitochondria the whole cycle normally goes through and phosphorylation is an important by-product of these reactions. This process of phosphorylation results in the production of high-energy phosphate esters, such as ATP.

Since ATP is one of the principal energy containing compounds of the body, it is very important in the whole energy cycle of the cell. It is, for example, the main source of energy in muscular contraction and for many of the synthetic processes of other cells. This briefly is the story of carbohydrate metabolism in the cell. What we want to try to do now is to demonstrate where this complex of activity is situated in the living cell. It appears that most of the processes of respiration and glycolysis actually take place in the cytoplasm and mitochondria.

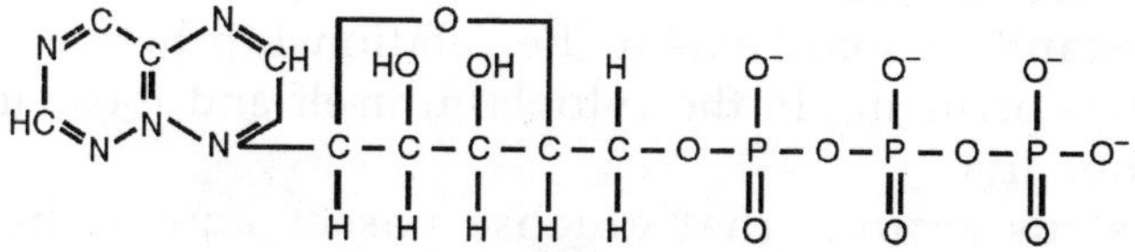

Fig. Adenosine Triphosphate (ATP)

Some years ago, the present author and R. J. Allen demonstrated that zymohexase which is really a complex of two enzymes and is concerned with the splitting of hexose diphosphate (the aldolase reaction) was localized in the cytoplasm of the cell and in the case of muscle fibers in between the fibrils in the sarcoplasmic material rather than in any of the formed elements. It is noted too that the results of differential centrifugation of cell homogenates have demonstrated that most of the enzymes responsible for the glycolytic cycle have been found to be present either in the supernatant or in "microsomes" and those concerned with the cytochrome system and Krebs cycle ire localized specifically in the mitochondria.

Now, since the microsomes mostly represent fragments of the endoplasmic reticulum, we can assume that most of the enzymes found in them can also be found in the membranes of the endoplasmic reticulum. In addition to the aldolase complex which has been demonstrated to be present in the supernatant, it has been found that phosphorylase and phosphoglucomutase and glucose 6-phosphate dehydrogenase have also been found in the supernatant and glycolysis has in fact been found to take place in this fluid—a fact which confirms that all the glycolytic enzymes necessary for this process must be present there.

However, glucose-6-phosphatase has been found not to be localized at all in the supernatant but exclusively in the microsomes (fragmented endoplasmic reticulum). There is some evidence too that hexokinase is present in the microsomes. On the other hand, DPNH and DPN and TPN-linked cytochrome c reductases which are found in high concentration in mitochondria have also been found to be present in the microsomes. It might be possible to suggest from these facts a tentative scheme which would help us to understand to some extent the relationship between the reactions occurring in the cytoplasm itself and those in the mitochondria.

Let us assume that glucose passes across the cell membrane, perhaps it is taken in by pinocytosis as has been suggested by a number of authors or, possibly, if and when the membranes of the reticulum are continuous with the outside of the cell, it may simply pass up the cisternae between the membranes of the reticulum and enter into the cell through the membranes of this reticulum. Let us assume that the glucose has passed into the cell by the process of pinocytosis. Then, with hexokinase present in the cytoplasm, it can be converted into glucose-6-phosphate and subsequently run through the cycle to pyruvate.

At this point (another system which we will discuss in a minute) associated with the mitochondria comes into play. If the glucose has to pass across the cell membrane or across the membranes of the endoplasmic reticulum, it should also be able to do this without too much difficulty. If any glucose-6-phospate should occur in the cisternae it will probably be able to pass freely into the cytoplasm because of the presence of glucose-6-phosphatase in the membrane. It may be that there is hexokinase in the fluid within the cisternae and that, perhaps, all the glucose there (if any) is first phosphorylated and then released in a timed fashion into the cytoplasm through the action of glucose-6-phosphatase in the membranes.

Perhaps this is a way of controlling the feeding of the carbohydrate fuel into the furnace. It will be remembered that glucose-6-phosphate passes cell membranes with difficulty unless there is the appropriate hydrolytic enzyme on the membrane. The glucose, once into the cytoplasm by whatever route, can be synthesized into glycogen or it can be brought to pyruvate by the enzymes of the glycolytic cycle which are known to occur in the cytoplasm. Once pyruvate is prepared, it is changed into acetyl coenzyme A and is ready for the next stage. For this it has to come into contact with the mitochondrial membrane. Whether it has to pass through the mitochondrial membrane and come in contact with the cristae we do not know, but the existing evidence suggests that the envelope and cristae of mitochondria have fundamentally the same enzymatic structure. Furthermore, there are pores and even large openings in some mitochondrial envelopes and there is thus little difficulty in metabolites passing in or out.

Now, it has been mentioned that a great proportion of the enzymes concerned with the Krebs tricarboxylic acid cycle are localized in the mitochondrial membranes and since acetyl coenzyme A appears to be the most commonly used fuel of this system, it is at this point that the mitochondria largely take over the final oxidative stages. Thus, as far as we can tell at the moment either in the cytoplasm of the cell itself or in the endoplasmic reticulum, glycolysis occurs with the production of either acetyl coenzyme A or pyruvic acid. We do not know at what stage the latter are fed to the mitochondria.

They must go largely to the mitochondria because these organelles contain the greatest proportion of the Krebs cycle enzymes—it is these latter which complete the oxidative breakdown of carbohydrate that was set into motion by hexokinase. The rate of penetration of these fuels into, or their absorption onto the mitochondria will be conditioned both by the nature of the mitochondria and the total area of mitochondria available. If under normal conditions we have a large number of mitochondria occupying the cell, we can assume, other things being equal, that oxidation should be proceeding at a fast rate.

However, where one gets a large increase of number of mitochondria as, for example, in starvation this may be compensatory hypertrophy and not indicative of increase in oxidative activity. We do not know whether the nature of the mitochondrial surface varies from time to time but we know that the surface area varies. Mitochondria, for instance, fragment from the filamentous and rodlike condition to the granular state under a variety of circumstances. These include mechanical damage to the cells, rough handling, influence of bacteria and other toxins, as a result of anesthesia and anoxia and so on. It can be shown mathematically that there is a greater surface area available the more fragmented the mitochondria become. Because of this either the pyruvate and/ or acetyl coenzyme

A can feed more rapidly into or become attached in larger amounts to the mitochondria when they are in a fragmented condition simply because there is a greater surface area. Thus the mitochondria in virtue of their ability to fragment and reform into the filamentous condition can act as a throttle controlling the rate of aerobic metabolism in the cell. They may also have an additional means of doing this, a sort of fine control. Since it is fairly certain that enzyme molecules are aligned along the cristae, these structures represent another surface where reactions can take place and reduction in the size and number of cristae would also have a throttling effect on the metabolism or synthesis due to mitochondria.

This can be seen in operation in the case of a mitochondrion which has accumulated a good deal of fat or other product of chemical reactions. In such a case, the cristae are reduced or absent altogether as if metabolic processes are brought to a virtual standstill because of the accumulation of reaction products. This brings us to another process of control, a chemical method known as "feedback" which will be discussed shortly.

Siekeiwitz believes that both glycogen synthesis and breakdown take place at the surface between the cytoplasm and the ergastoplasmic membranes in association with the enzymes present at those surfaces. He points out, however, that only glucose-6-phosphatase and DPNH and TPNH cytochrome c reductases (these latter are believed to act as coenzymes for glycolysis in the early stages of glucose oxidation) have been found to be associated with the microsome fraction, but he believes that the site of effective action of the other enzymes might be at the interface between the membranes and the matrix of the cytoplasm. He suggests that hexokinase might be activated at the membrane surface of the endoplasmic reticulum. Other cofactors such as glucose-1-6-diphosphate and adenosine monophosphate possibly also bind their appropriate enzymes to the E.R. membranes.

Siekewitz has also discussed certain biochemical aspects of the control of glucose metabolism. First of all he points out that, if the concentration of glucose in the lumen of the endoplasmic reticulum is in equilibrium with the glucose concentration in the blood (this assumes at least temporary continuity between the lumen or cisternae of the E.R. and the exterior of the cell), there exists then a mechanism whereby the glucose level in the blood would definitely affect intracellular glucose equilibrium.

Thus a reduced production of glucose from the diet would lead to reduction of glucose in the blood and this would cause a reduction of the amount of glucose in the fluid within the endoplasmic reticulum. The latter result would lead to increased phosphatase activity which would cause an increase in the breakdown of glycogen (e.g., in the liver cells) and a production of glucose which would pass out from the endoplasmic reticulum into the blood stream.

Siekewitz points out that hexokinase, phosphoglu comutase and phosphorylase might have their activity enhanced if they were attached to the endoplasmic reticulum membranes. In the case of phosphorylase, active phosphorylase B has to undergo a conversion to active phosphorylase A before it can have any catalytic effect. An enzyme phosphorylase B-kinase carries out this conversion by phosphorylating the enzyme in the presence of ATP. Siekewitz suggests that this kinase may be part of the membrane of the endoplasmic reticulum and that the activating process for phosphatase in the cell might consist of moving it out of the cytoplasm onto the site of the E.R. membranes.

Siekewitz points out that it is possible that hormones control this type of movement of enzymes within the cell and its internal membranes. At this point one might remember the work quoted earlier by Brandes and the present author in which it was demonstrated that shifts of acid phosphatase activity took place from the Golgi apparatus to the nucleus in ventral-lobe prostate cells following castration and that acid phosphatase activity was restored to the Golgi apparatus following implantation of the male sex hormone. Also of interest are the further studies of Brandes in which he has demonstrated that in castrated animals there is a rearrangement of the membranes of the endoplasmic reticulum. These studies demonstrate a definite morphologico biochemical effect on the part of the male sex hormone. Siekewitz suggests that the hormones concerned with carbohydrate metabolism do not act directly on the enzyme as such but bring it and the substrate and various cofactors together at a suitable surface and then complex them together there. Other enzymes may be localized on the endoplasmic reticulum e.g. 5 nucleotidase. The nucleus, as will be described later, appears to be enclosed in a fold of the double membrane of the endoplasmic reticulum and not to have a membrane of its own. Since the E.R. spaces might be connected directly to the exterior of the cell, it is possible that glucose coming from outside the cell could be converted into glucose-6-phosphate in the E.R. lumen and thus be prevented from passing through the E.R. membrane.

Thus it could pass along all the ramifications of the E.R. canals and come in direct contact with the nuclear fold of the reticulum. If this membrane contains glucose-6-phosphatase, it could penetrate through into the nucleus without having to pass through the cytoplasm at all. Histochemical preparations show that the glucose-6-phosphatase reaction in a particular histological section is not always positive for all nuclei in the section and it is of interest that Siekewitz has pointed out that enzymes may be present or activated at the E.R. membrane only when the glucose concentration reaches a critical level.

It is of interest that histochemical studies of the cells of many organs demonstrate the fact that many dephospho rylating enzymes as well as glucose-6-phosphatase show an association with nuclear membranes—perhaps here lies the mechanism whereby even low levels of glucose-6-phosphate could penetrate readily through into the interior of the nucleus. The endoplasmic reticulum could provide a pathway straight to the nucleus which would prevent glucose from getting in contact with or passing through the cytoplasm where it could be attacked by glycolytic enzymes. This may be the mechanism by which a supply of glucose to the nucleus is ensured. Glucose could also be supplied to the nucleus from the cytoplasm by the process of glycolysis.

In this case, glucose being formed from glycogen as glucose6-phosphate would be dephosphorylated and pass through the E.R. membrane and into the lumen where possibly it would be rephosphorylated to prevent its passing back again and would thus move along the lumen to the nucleus. Hence the nucleus could get its glucose directly from the cytoplasm via the endoplasmic reticulum or directly from the outside (the latter, however, only if a direct connection really exists).

In many cells the mitochondria can also be seen to be completely surrounded by endoplasmic reticulum. It is possible that the mitochondria themselves obtain their glycolytic fuel directly as a result of glucose passing through the E.R. membranes undergoing glycolysis there and the glycolytic products feeding directly to the mitochondria.

The control of the rate of different types of metabolism, particularly respiration, in the cell can depend on two main factors. First, a structural factor which brings into apposition the appropriate reactants and which varies with the extent of the surfaces available for these processes to take place and, second, it may also depend on some chemical feedback where an excessive production of one kind of compound inhibits its continued production or slows down further synthesis. Alternatively, the metabolism in any one particular direction may be affected by the absence of a limiting amount of some specific substance or compound in the reaction chain.

Sir Hans Krebs in a recent article entitled *"Rate Limiting Factors in Cell Respiration"* discussed the control of energy utilization and pointed out that, in unicellular organisms, energy can be obtained directly from oxidation if air is present but if air is not present then anaerobic fermentation takes its place and energy is obtained by this source. In this instance it is the supply of air or oxygen which regulates which of these mechanisms comes into use and oxygen is, in fact, the "rate limiting factor." In higher animals the ability to undergo fermentation or an aerobic oxidation is still present and can be particularly well demonstrated in muscle. Krebs pointed out that the chemical systems in the cell which are concerned with the function of regulation are all fairly simple reactions but that there is an elaborate interlocking of these reactions. By this he means that the individual reactants may take part not only in more than one reaction but in very many different processes.

One of the difficulties in sorting out such a complex of activity is that not all the component reactions of this elaborate interlocking series are known and we are dealing with a heterogeneous system in which there are many varied membranes and different spatial arrangements of the various reactants. He also points out that regulation is probably a matter of reaction velocities, some of which will be accelerated and some slowed down and the question that has to be decided is the degree to which any of these are ratelimiting. Krebs illustrates this point by considering the amount of oxygen used by 4-ml. sheep heart homogenate, which contained about 10% of tissue.

Thus one can demonstrate that the addition of glycogen adds nothing to the oxygen uptake so the amount of glycogen present is not a limiting factor in this system. The same applies if glucose is added instead of glycogen. On the other hand, when acetate, pyruvate, or other intermediates of the tricarboxylic acid cycle are added there is an appreciable increase in the rate of oxygen uptake. The fact that the oxygen uptake in this system can be increased if suitable substrates are added to the mixture demonstrates that the electron transport system from DPNH to oxygen is not being used to its full capacity, thus it cannot be the factor which is limiting the uptake of oxygen. Certain special substrates which are known to reduce either DPN or flavoprotein seem to be able to increase oxygen consumption.

Thus the limiting factor appears to be that, the mechanism for the transport of hydrogen from DPN or flavoprotein is not being used to its full capacity if these special substrates are not present. The limiting step therefore is really the first stage in the electron transport system. If it is found that a particular substrate increases the rate of respiration this is due to the fact that the substrate reacts more readily or easily with DPN or flavoprotein than any endogenous substrate already present. This is the reason why pyruvate or á-ketoglutarate or succinate are responsible for the stimulation of respiratory rate in the experiment quoted. However, even if we accept this we are still faced with the identification of the factor that decides the rate of reaction between substrate and DPN or flavoprotein.

Studies with dinitrophenol which decouples oxidative phosphorylation from respiration is of interest and helps to throw light on this problem. Perhaps we should first say a word or two about this action of dinitrophenol. The uncoupling of oxidative phosphorylation from respiration has been compared to putting a car into neutral gear and still leaving the engine running. If the respiratory activities are regarded as the engine and the phosphorylation as the process of making the car go, then dinitrophenol uncouples the engine from the transmission of the car, the engine continues to turn but the car does not move; in the cell the respiration goes on quite happily but no ATP is formed. Normally ATP is formed from ADP and inorganic phosphate and thus the factor which limits the rate of oxygen consumption and oxidation of pyruvate in the normal system such as we have described is not really the amount of enzymes present but actually the level of either ADP or inorganic phosphate.

In the experiments carried out by Krebs, inorganic phosphate was present in a fairly good concentration and further quantities added to the system did not stimulate respiration. Therefore it is almost certain that the limiting factor must be the amount of ADP which is available. These experiments demonstrate the type of chemical control which a single compound can exert on a whole chain of reactions. One should also remember that another mechanism, a structural one that controls the rate of respiration, is the rate at which glucose can enter the cell and this can be hormonally controlled, although the method of action of the hormone is not exactly known.

It is of interest that one of the factors which probably affects the rate at which glucose can enter the cell (if, in fact, the endoplasmic reticulum is continuous with the outside of the cell) is the degree of complexity of the endoplasmic reticulum. If this structure develops many ramifications, as presumably it seems able to do in certain cells such as spermatocytes, as demonstrated by Fawcett, then the surface area available for glucose to enter into the cytoplasm of the cell and so be metabolized is enormously increased or, conversely, it may be decreased by a reduction in complexity of the reticulum. To return to the subject of respiration and ATP formation, the reaction for this process can be given as follows:

$$C_6H_{12}O_6 + 6\ O_2 + 38\ ADP + 38\ H_3PO_4 \rightarrow 6\ CO_2 + 44\ H_2O + 38\ ATP$$

This is the general reaction and is a summary of all the complex intermediatary reactions which in the end simply produce carbon dioxide, water and ATP. Since, as Slater and Houlsman have pointed out, cells contain only relatively small amounts of ADP, as soon as it is all converted into ATP the process of respiration will stop—ADP is thus the limiting factor. However, when the cell is stimulated to do work there is a breakdown of ATP according to the formula given by Slater and Houlsman:

$$38\ ATP + 38\ H_2O \rightarrow 38\ ADP + 38\ H_3PO_4 \rightarrow \text{work}$$

Respiration can go on so long as there is some left to be resynthesized into ATP. The addition of further ATP will, of course, keep respiration going.

It is of interest that, if mitochondria that have been separated from the cell by differential centrifugation are permitted to stand for some hours at room temperature, the phenomenon of uncoupling (which can also be brought about by dinitrophenol) of the oxidative phosphorolytic system from respiration takes place. This type of mitochondrial preparation is described as "aging" mitochondria and it is tempting to speculate whether in senescing tissues there may not be a progressive uncoupling of respiration from oxidative phosphorylation or a progressive hydrolysis of ADP so that less and less of this becomes available for synthesis of ATP. It has been possible to isolate from mitochondria which have been aged in this way, a heme compound which will actually produce this uncoupling reaction. It has been described and given the name "mitochrome" and is fundamentally a pigment.

However, there appears to be a lipid component in this "mitochrome" heme-protein preparation which is the actual factor responsible for uncoupling and the heme protein of the mitochrome is not, in fact, the uncoupling factor at all. Mitochrome is very similar in structure and form to cytochrome and is probably derived from it. Certain unsaturated fatty acids such as oleic acid are also found to be active as uncoupling agents and the lipid isolated from the mitochrome particle also appears to contain an unsaturated fatty acid. The fact that an uncoupling agent can be produced *in vitro* this way is of considerable importance since it seems possible that the formation of such a substance in mitochondria might take place in vivo and may itself function as a controlling agent for respiration and oxidation.

FAT METABOLISM

The general view of metabolism of fatty acids at present is that they are broken down beginning at the end of the chain where the carboxyl group is and then the chain is progressively degraded as two carbon pieces are removed by a process of oxidation.

This type of oxidation of fatty acids is known as ß-oxidation and is so called because the fatty acid is attacked oxidatively at the ß-carbon atom in the first instance. Among the products of this oxidation is the formation of acetyl coenzyme A and acyl coenzyme A. The latter can be subjected to further oxidation with the production of more acetyl coenzyme A (CoA).

Although very little has been said about the mechanism of degradation of all these fatty acids, it is of interest that the enzymes which catalyze these reactions are all located in or on the mitochondria. The acetyl CoA can enter the Krebs cycle by condensing with oxaloacetic acid and the final oxidation thus follows the same path as the carbohydrates. The acetyl CoA formed from pyruvic acid (i.e., from carbohydrate breakdown) can be used to synthesize fats and likewise so can acetyl CoA produced as a result of protein metabolism. We see, therefore the reason why the Krebs tricarboxylic acid cycle has been spoken of as the meeting place of protein, fat and carbohydrate metabolism.

Where precisely in the mitochondria the enzymes responsible for ß-oxidation of fatty acids are centered is not known for certain. It is very likely that they are more associated with the outer membrane of the mitochondrion than with the cristae so that the problem of penetration of the fatty acid through the membrane of the mitochondrion does not become so important. On the other hand, there is some evidence that pores occur in the mitochondrial membrane and if this is the case it is possible for the long-chain fatty acids to pass into the interior of the mitochondria and to become subject to ß-oxidation at this site by enzymes located in the cristae.

Hoberman has shown that in the mitochondria, deuterium labeled DPNH, which has been reduced during the oxidation of fat in the organelles, is not available for reactions that take place in the outside cytoplasm. This suggests that the enzymes concerned with the oxidation are localized within the cristae. It is of interest that recent electron micrographs of the adrenal cortex have shown large areas of the surface where the mitochondrial membrane is incomplete and the interior of the organelle is thus open to the penetration of the largest molecules and even particles of fat. Novikoff's scheme for the differential distribution of biochemical activities in liver cells.

Fat is metabolized largely in the way already described, but the synthesis of fat is also an important part of the activity of the cell. Fat cells have an important mechanical function to perform and fat itself is a valuable reserve store of energy. Fats, fatty acids and phosphorylated fats (phospholipids such as lecithin) are also important structural units of cell, mitochondrial and other membranes and their synthesis becomes an important cellular activity. Fatty acids are made up of long chains of carbon atoms, usually an even number, with a COOH group stuck at one end. Fat is formed from a fatty acid molecule by a combination between the latter and a molecule of alcohol such as glycerol. The link takes place through the COOH group which reacts with the OH of the alcohol to eliminate a molecule of water. Fatty acids may be short or long or intermediate chained, a typical short-chain fatty acid is acetic acid which has only 2 carbon atoms and a typical long-chain fatty acid is palmitic acid which has 16 carbon atoms.

In the synthesis of these long-chain fatty acids, the starting point appears to be acetic acid. This combines with CoA to form acetyl CoA. This condenses with CO_2 to form malonyl CoA (the coenzyme A ester of malonic acid). This latter compound then condenses with another molecule of acetyl CoA to form an intermediate substance which becomes reduced to form a 4-carbon fatty acid which condenses with a molecule of malonyl CoA to give a 6-carbon fatty acid and so the carbon chain is built up. Then, as described, combination of the appropriate fatty acid with an alcohol such as glycerol gives an ester known as a fat.

Phospholipins such as lecithin, which play a very important part in the structure of the cell membranes, also have to be synthesized by the cell. These compounds are not only esters of a fatty acid but are simultaneously esters of phosphoric acid. In this synthetic activity ATP plays an important part. It starts the ball rolling by phosphorylating one of the OH groups of glycerol. To the other two OH groups, CoA esters of palmitic acid are attached. At this point the phosphoric acid group that was put on in the first step drops off. The molecule then reacts with a compound known as cytidine diphosophocholine. This compound, which is really a coenzyme, drops the diphosphocholine part of its molecule nicely into the spot on the OH group which had been vacated by the first phosphoric group and so the lecithin is formed. Other coenzymes are concerned with the production of the other steps in the synthesis but these will have to be studied in more specialized works of biochemistry. Dr. D. E. Green has pointed out that one of the surprising things about the synthesis of fatty acids is that the synthesis stops at 16-carbons. It is rare to get 12- and 14-carbon chain fatty acids and 18- or more carbon chains scarcely ever form. What tells the cell to break off the synthesis at that point is certainly an intriguing problem.

Although one would expect that mitochondria would be the principal fatty acid synthesizers of the cell, the belief at the moment is that they are not the site of synthesis and that it takes place at the surface of the endoplasmic reticulum. However, in the production of fats from the fatty acids and especially in the case of lecithin synthesis where ATP is required, the mitochondria make a contribution to the synthesis because of their ability to produce the latter material. Mitochondria do play a direct role in fat metabolism but, this role as mentioned earlier in this chapter, appears to be in fat degradation rather than in fat synthesis.

PLANT CELLS

So far we have considered only animal cells and it is of interest that Hackett in the *International Review of Cytology,* has pointed out that in plant cells glycolysis involves the plastids, the soluble fraction of cells and possibly the nucleus as well as the mitochondrial enzymes. He points out that many enzymes which are involved in the Krebs cycle are not exclusive to the mitochondria and that hydrogen transfer is not confined to these organelles.

Mitochondria he says react with the nucleus in the process of phosphorylation, they react with the chloroplasts in photosynthesis and they react with the microsomes in protein synthesis in the endoplasmic reticulum. He believes that a close relation between the cell membrane and the mitochondria may play an active part in the movement of substances into the cell or possibly in the growth of the cell wall of plants.

Furthermore, he reminds us that the real unit is the cell itself, its various parts work as an integrated unit and one should only break them down for the purpose of trying to analyse the various processes in which they participate.

MITOCHONDRIA: POWER PLANT OF CELL

For thousands of years, men puzzled over the question of where body heat comes from. This heat was recognized as nearly synonymous with life itself -- but what produced it? The source of the energy of man or beast was equally mysterious. In recent decades it has become clear that both heat and energy can be traced to some sausage-shaped organelles in our cells, the mitochondria.

Of course the ultimate source of energy for all plants and animals is sunlight. But the sun's energy can be harnessed by plants, through photosynthesis and stored in molecules of carbohydrates. When animals eat these carbohydrates and break them down to carbon dioxide and water, with the help of oxygen and an arsenal of enzymes, large amounts of energy become available. Animals immediately convert this energy into molecules of high-energy ATP (adenosine triphosphate) -- the universal currency of energy in living things. Excluding only the very first stages in carbohydrate breakdown, which are called glycolysis, the entire, complicated process of energy transfer to ATP takes place within the mitochondria.

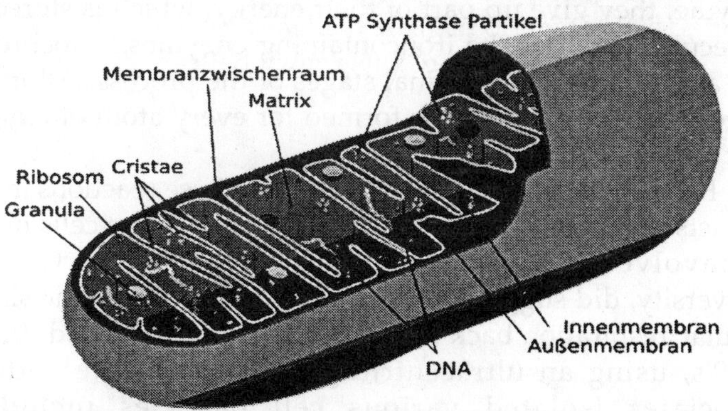

Fig. A Mitochondrion

Mitochondria are the largest organelles in the cell, after the nucleus, yet some cells have more than a thousand of them. Thin and long, they vary in diameter from 0.5 to 1 micrometer and in length up to 7 micrometers. Their thread-like outlines can be seen with a good light microscope. Although they were first observed in 1850, it took a century to visualize their internal structure and understand their function.

Actually the story goes back even further, to the 18th century, when Lavoisier showed that animals require oxygen for respiration. This led chemists to study cellular respiration, the process by which living cells use oxygen to release the chemical energy stored in foodstuffs. Early in the 20th century biochemists discovered that these reactions fall into two major groups:

- The carbon pathway, where a series of chemical reactions, each requiring a specific enzyme, breaks down the carbohydrates into carbon dioxide and hydrogen;
- The hydrogen pathway, which transfers the hydrogen to oxyen in stages, forming water and releasing energy.

The whole process is organized much like an assembly line. In the hydrogen pathway, the hydrogen's electrons pass through an "electron transport chain" made up of enzymes that act as carriers and as the electrons move from enzyme to enzyme, they give up part of their energy, which is stored in molecules of ATP. The ironcontaining enzymes, cytochrome a, b and c, carry out the final stages of the process. All in all, three molecules of ATP are formed for every atom of oxygen that is used up in respiration.

For a long time, biochemists studied these reactions in cell extracts without worrying about what parts of the cells might be involved. One scientist, B.F. Kingsbury of Cornell University, did suggest that mitochondria might be the site of cellular respiration back in 1912, but this was ignored. In the 1940's, using an ultracentrifuge, Albert Claude and his associates isolated various cell particles including mitochondria. Shortly afterwards, in 1948, Albert Lehninger of Johns Hopkins and Eugene Kennedy of Harvard showed that the reactions leading to the synthesis of ATP occurred in the mitochondria.

In the early 1950's, Palade and the Swedish scientist Fritiof Sjostrand reported that mitochondria are bounded by a membrane and that they have a system of parallel, regularly spaced inner ridges which were named cristae. It is now clear that there are, in fact, two membranes around the mitochondria of plants, protozoa, molluscs, insects and man: an outer membrane, set off by a fluidfilled gap; and an inner membrane which is folded inward at various points to increase its surface, forming the cristae, so that the enzyme molecules of the electron transport chain, which are attached to this inner surface in specific sequences, can be packed more densely side by side in the mitochondria.

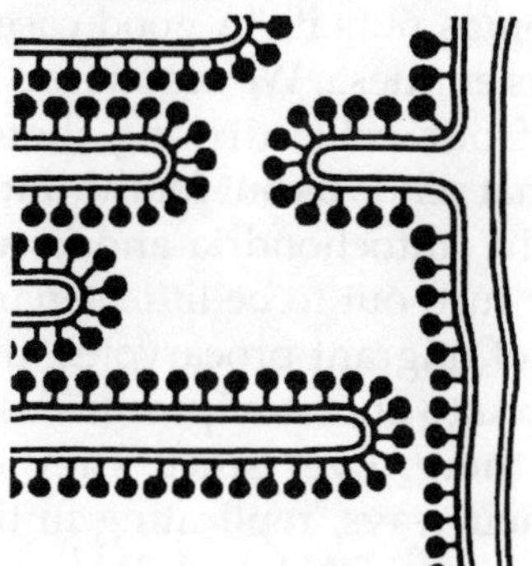

Fig. The Intricate Internal Structure of a Mitochondrion, with Outer Membrane and Particles on the Folded Inner Membrane, Part of which forms the Cristae

This general design for mitochondria seems to have existed unchanged, through the evolution of plants and animals, for more than 1.2 billion years. Though mitochondria are generally visualized as ovalshaped, they can change shape quite drastically. They swell and contract in response to various hormones and drugs. Even in very low concentrations, Lehninger found, the thyroid hormone thyroxin will make mitochondria swell and will simultaneously "uncouple" the electron transport chain from the synthesis of ATP, so that even though carbohydrates may be metabolized and oxygen consumed, no ATP will be manufactured. ATP, on the other hand, makes mitochondria contract. This swelling and contraction appear related to the movement of water in, out and through cells, which is of particular importance in the kidney.

The more scientists learn about mitochondria, however, the more their curiosity is aroused, for these are without a doubt the strangest of the cell's organelles. To begin with, mitochondria contain their own hereditary material (DNA), which resembles that of bacteria far more than that of animal cell nuclei. Because of this similarity, many scientists believe that mitochondria are derived from bacteria that infected a primitive cell and evolved into a useful symbiotic relationship with it.

As Lewis Thomas puts it, "A good case can be made for our nonexistence as entities... We are shared, rented, occupied. At the interior of our cells, driving them, providing the oxidative energy that sends us out for the improvement of each shining day, are the mitochondria and in a strict sense they are not ours. They turn out to be little separate creatures, the colonial posterity of migrant procaryotes, probably primitive bacteria that swam into ancestral precursors of our eucaryotic cells and stayed there. Ever since, they have maintained themselves and their ways, replicating in their own fashion, privately, with their own DNA and RNA quite different from ours. They are as much symbionts as the rhizobial bacteria in the roots of beans. Without them, we would not move a muscle, drum a finger, think a thought."

"My mitochondria comprise a very large proportion of me. Looked at in this way, I could be taken for very large, motile colony of respiring bacteria, operating a complex system of nuclei, microtubules and neurons for the pleasure and sustenance of their families and running, at the moment, a typewriter," continues Thomas somewhat facetiously. "I am...obliged to do a great deal of essential work for my mitochondria. My nuclei code out the outer membranes of each and a good many of the enzymes attached to the cristae must be synthesized by me. Each of them, by all accounts, makes only enough of its own materials to get along on and the rest must come from me. And I am the one who has to do the worrying."

During the past few years, scientists have begun to learn which proteins and enzymes of the mitochondria are synthesized according to directions in the mitochondrial DNA and which are controlled by the genes in the nucleus. This information would be of considerable importance for the design of antibiotics and drugs, as well as for the understanding of genetic diseases. For example, researchers have found that chloramphenicol -- a powerful antibiotic which can cause a fatal anemia -- interferes with the synthesis of mitochondrial DNA, while other antibiotics do not.

Mitochondria are just as important to the life of each cell as the heart is to the human body. At least 95 percent of our cells' energy comes from mitochondria. (Only the mammalian red blood cells, which have no nuclei, get along without mitochondria; they obtain whatever energy they need from the partial breakdown of glucose in their cytoplasm). Growing cells continuously synthesize nucleic acids, proteins and other components, for which they need the energy that is stored in molecules of ATP. Muscle cells which carry out mechanical work, plant cells which pump water against the force of gravity and many other cells also require a continuous source of energy.

Besides supplying this energy, mitochondria apparently help to control the concentration of water, calcium and other ions in the cytoplasm. They are also deeply involved in the breakdown and recycling of sugars, fatty acids and amino acids to recover their energy, as well as in the formation of urea as a waste product, to be excreted through the kidney.

These strange organelles may thus hold clues to a variety of seemingly unrelated diseases. They have been linked to some liver diseases (e.g., viral hepatitis, obstructive jaundice and cirrhosis of the liver); some muscle diseases (e.g., familial myotrophic dystrophy); and some kidney diseases. In all these conditions, pathologists have seen large amounts of materials pile up inside the mitochondria as "inclusions" of varying shapes and sizes, presumably waiting to be processed. What these inclusions consist of remains a mystery, but if scientists could find out, they might discover which specific components (for instance, enzymes) are involved in each disease.

In addition, several calcium disorders may be related to mitochondrial defects. For example, abnormal concentrations of calcium (possibly caused by malfunctioning mitochondria) may lead to arthritis, bursitis, or arterial disease. While all these leads are promising, the greatest spinoff from research on mitochondria may be in the area of energy -- not only for the cell, but for general use by mankind.

As the oil crisis has made us realise, we presently depend on a limited supply of fossil fuels -- coal and oil -which are of biological origin but cannot be replenished. On the other hand, nature contains a far greater resource: a molecular know-how and technology gained over millennia, by which biological systems manufacture ATP from solar energy. This technology awaits us. It exists in very similar, if not identical forms in mitochondria and in their mirror image, the chloroplasts, which are found in green plants, where they do the work of photosynthesis, converting the sun's energy into carbohydrates. Like mitochondria, chloroplasts have an outer and an inner membrane. They have some of their own DNA. They, too, may have originated as some kind of infection, probably with blue-green algae. And it is on chloroplasts that we ultimately depend for nearly all our energy -- that of our bodies as well as that which is stored in deposits of oil and coal.

If scientists ever unravel the details of the process by which chloroplasts store solar energy in carbohydrates and mitochondria release it, we may be able to produce unlimited quantities of usable energy.

MICROTUBULES, THE CELL'S PHYSICAL PROPS

All the organelles described so far are bound by membranes. For a long time, the rest of the cytoplasm -- a liquid called the cell sap, or ground substance -- appeared totally unstructured. But as scientists learned to use newer and gentler fixatives to prepare cells for electron miscroscopy, more and more tiny structures materialized in this soup.

The most prominent of the new structures are two organelles which provide an intricate system of physical support for the cell -- the equivalent of buttresses or skeletal bones -- as well as a contractile mechanism for cell movement: microtubules and microfilaments. Neither is bound by a membrane.

Microtubules were first noticed in the middle 1950's, but they were seen only rarely until the development of glutaraldehyde as a gentle fixative in 1963. Extremely thin cylinders, about 200 to 300 angstroms in diameter and of variable length, microtubules are constructed chiefly of proteins called tubulins. In 1967, Edwin Taylor and his associates at the University of Chicago discovered that colchicine -- a chemical used to arrest cell division -- did its work by binding to the protein tubulin. This pointed to the role of microtubules in mitosis and helped to answer a question which had worried generations of biologists: How do chromosomes sort themselves out into two sets and separate during cell division? The details of mitosis would have been very hard to explain without understanding the chemistry of the spindle fibers which organize this separation.

In fact, if there were no microtubules, scientists would have had to invent them. Something of this sort was needed to account not only for mitosis, but for the surprisingly firm structure of such seemingly vulnerable fibers as the long, thin axons of nerve cells, as well as for the unique structure of some red blood cells, which are held in a disc-like shape by hoops composed of microtubules.

Microtubules are also involved in assisting the transport of substances in and out of the cell and in the motion of cells themselves. They make up the cilia and flagella, whip-like filaments which project from the surface of cells and move rhythmically. Large numbers of cilia are found on cells that line the respiratory system, for instance, where they help to sweep out dust and debris. Both cilia and flagella play an important role in human reproduction: the coordinated beating of cilia in the oviduct produces a sort of current which draws the female's egg into the uterus, while the rapidly thrashing tail of sperm actually is a flagellum.

MICROFILAMENTS

The movement of living cells has fascinated biologists ever since Leeuwenhoek discovered "tiny animalcules prettily a-moving" under his simple microscope. But for a long time, scientists' efforts to study such movements were stymied by techniques which required killing cells before they could be examined under the microscope. It is only since the development of time-lapse photography, together with the phase-contrast and interference microscopes, that researchers have been able to study cells while they were moving.

There seem to be two main kinds of cell motion: That produced by the rhythmic beating of special structures on the cell surface (such as cilia and flagella) and that connected with some wispy organelles inside the cell, microfilaments. The action of these microfilaments allows cells to "crawl" along surfaces by forming extensions, called pseudopods, towards which the bulk of the cytoplasm flows. While the precise nature of this movement is not understood, it seems to involve the continual transformation of the liquid parts of the cytoplasm (the cell sap, or ground substance) into a viscous gel with the help of calcium and its subsequent dissolution into a liquid again. Amoebae, white blood cells, macrophages and the tips of growing nerve cells "crawl" in this fashion. So, probably, do cancer cells.

Microfilaments apparently consist of thin strands of actin, a protein. Actin has been studied for nearly half a century in muscle cells. In the fifties scientists learned, primarily through the work of Hugh Huxley at the Medical Research Council laboratories in England, that contraction in the skeletal muscles is produced by the sliding of thin filaments of actin between thick filaments of another protein, myosin. But it is only in the past 15 years that researchers have found evidence of similar actin-myosin interactions, involving microfilaments, in many other kinds of cells. The whole process of secretion, as when cells discharge hormones, enzymes or other proteins, may depend on such interaction. So may the movement of many cells which help to defend the body against invading organisms.

As the President's Biomedical Research Panel put it recently, "Here again is a striking example of biological versatility and -- with it -- the promise of a common key to the solution of many problems in basic or applied biomedical sciences." Current research on all the cell's organelles -- studies of the nucleus, ribosomes, ER, Golgi, lysosomes, mitochondria, microtubules and microfilaments -- holds such promise. However the common key that may unlock the greatest number of health benefits may well be found in the cell's filmy membranes, particularly in the plasma membrane at the surface of the cell.

MEMBRANES, THE CELL'S TOUGH, DELICATE GUARDIANS

Whatever flash of lightning there was that organized purines, pyrimidines and amino acids into macromolecules capable of reproducing themselves, it would not have yielded cells but for the organizational trick afforded by the design of a membrane wrapping." He visualizes a kind of bubble in which the first macromolecules must have been enclosed, to protect them from being dissipated or dissolved in the strong salt of the early sea.

Membranes are now arousing enormous curiosity among biologists because it has become clear that these diaphanous films, less than 100 angstroms (1/100,000th of a millimeter) thick, control some of the most vital functions of the living cell. Furthermore, there are layers upon layers of them, in and around various organelles.

The double-membrane envelope which forms a channel around the nucleus not only protects the genetic material but also connects with the "plumbing" the system of membranous tubes, folds and compartments -which makes up the rough and smooth endoplasmic reticulum and the Golgi apparatus.

The membranecovered mitochondria contain a separate inner membrane system of their own. And at the surface of the cell, the all-powerful plasma membrane -- an organelle in its own right -- stands between the cell and the world.

Like the skin that covers our bodies, the plasma membrane creates a separate environment within which the biochemical processes of life can take place. In addition, it controls everything that goes in or out of the cell. It also receives signals from other cells, interacts with them to form tissue, recognizes attacking viruses, responds to hormones. And as a result of its contact with the outer environment, it probably tells the nucleus when to divide -- or when to stop. Despite its fragility, it is a tough and resourceful guardian of the cell's interior.

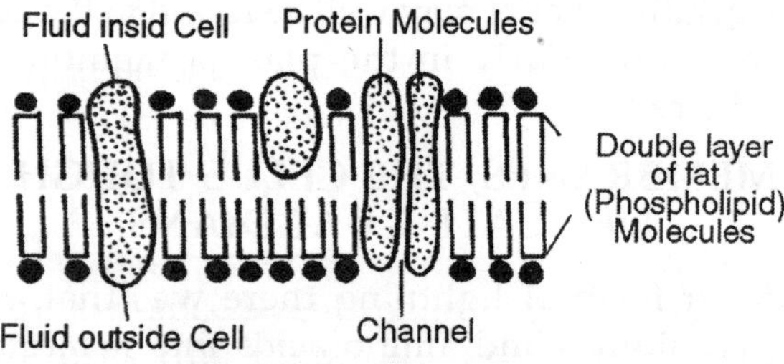

Fig. The Plasma Membrane

The secret of the membranes' power lies in their ingenious composition and structure. After years of puzzlement and speculation, biologists now have some idea of how these thin membranes are organized. Of course there is no "typical" membrane, since the various types that exist within each cell and in different kinds of cells differ in composition and function. Furthermore, any particular membrane is likely to change from moment to moment, in response to its environment. Nevertheless all cell membranes follow the same general pattern.

Their first task is to provide effective compartments. Since cells live in a watery environment and also contain a large amount of watery fluid, in which the organelles float, anything which forms a compartment must be pretty water tight. In fact, membranes have been found to consist mostly of lipids (fat) interspersed with proteins and since fat and water don't mix, this creates an efficient barrier to keep foreign substances out and water in. Substances that dissolve well in lipids will not dissolve well in water and vice versa (excepting only detergents and gases, such as oxygen and nitrogen, which dissolve in both).

A membrane is not just a layer of fat, however, but an intricate, elastic and fluid "bilayer" consisting of two layers of phospholipids (each layer one molecule thick) interspersed with proteins which are either embedded in the lipid layers or loosely attached to their surface. The two layers of phospholipids arrange themselves very neatly, as described in 1938 by James Danielli, who was then at the State University of New York at Buffalo: The part of the phospholipid molecule which is water-loving (a small portion containing charged phosphate, which is electrostatically attracted to water) faces out towards the outside world, or towards the watery cytoplasm, while the rest of the molecule (consisting of hydrocarbons, which are insoluble in water) is pushed to the centre of the membrane.

Actually the layer of membrane which faces the outside world fulfills quite different functions from the layer which faces the cytoplasm. Therefore the two layers are not symmetrical; each consists of a different array of lipids and proteins.

MEMBRANE PROTEINS

In 1966, S.J. Singer of the University of California at San Diego proposed a "fluid-mosaic" model for cell membranes. He compared some of the proteins in cell membranes to "icebergs floating in a sea of lipids," and said that their movement would create an ever-changing "fluid-mosaic" pattern. In his view, most of these proteins are folded up so as to have one hydrophobic (waterhating) and one hydrophilic (water-loving) end -- just as do the phospholipid molecules, though of course the proteins are much larger. He also suggested that if the proteins are large enough to span the entire thickness of the membrane, they must have two hydrophilic ends (which protrude) and a hydrophobic middle (which is embedded in the phospholipid layer). Although such tri-partite proteins were unknown at the time, they have been found to exist in membranes.

The proteins in membranes are extremely important because they do most of the specialized work, while the lipids serve largely as a diffusion barrier. In the plasma membrane, for example, protein "receptors" respond to specific hormones, neurotransmitters and antibodies. They selectively accept specific chemicals from outside the cell, which must fit them as precisely as a key in a lock and send appropriate signals into the cell's interior.

The receptors on which so much depends must make exquisitely refined discriminations at every phase of the cell's life, beginning with fertilization. For instance, sperm cells recognize egg cells only when these belong to the same or to a very closely related species. As George Palade puts it, "Imagine what problems could arise for all the creatures that live in the sea, with all kinds of eggs and all kinds of sperm floating around in the water! Yet because of the precise information obtained through receptors on the cell membranes, fertilization occurs only within specific species." The intricate process of differentiation, through which embryonic cells take on specialized functions, also depends largely on specific cells' recognizing one another through the proteins in their plasma membranes.

The failure of protein receptors to do their job properly may lead to a variety of diseases. "We are beginning to understand that there may be either inherited or acquired defects in the organization of cell membranes," says Palade, who now directs the Centre for Research on Cellular Membranes which NIGMS started at Yale University in 1974. "For example, an organism may produce enough of a hormone, such as parathormone or insulin, but not have enough receptors for it in the membrane of the target cell. This may look like a hormone deficiency, when in fact it is not. We are finding many membrane disturbances of this sort. Or there may be an excess of receptors on the membrane of the target cell, leading to a syndrome that mimics the overproduction of a hormone or neurotransmitter; for example, some forms of hypertension and cardiovascular disease may be produced by an excess of receptors for hormones on the membranes of smooth-muscle cells on the arterial walls."

PLASMA MEMBRANE

As primary guardian of the cell's gates, the plasma membrane is responsible for keeping the chemical composition of the cell's interior constant within very narrow limits. The interior of the cells of higher animals needs to be high in potassium ions and low in sodium ions, for instance, while their exterior environment -- blood and body fluids -- has the reverse concentration, being high in sodium (like sea water) and low in potassium. This creates a difference in electrical potential of about one-tenth of a volt or higher across the cell membrane, which is particularly important to the action of nerves and muscles. Scientists are only beginning to understand how the plasma membrane accomplishes such impressive feats.

The simplest traffic across the plasma membrane is passive transport, which requires no energy. The molecules diffuse through the membrane at rates determined by their solubility in lipids, their size and differences in their concentration. Those that are highly soluble in lipids move most rapidly. Water and other small molecules which are not soluble in lipids can also diffuse through the membrane, with the smallest ones passing through at the highest rate. (If there were no membrane at all, however, water could diffuse in or out of the cell 100,000 times more rapidly. It is believed that there are "channels" or "pores" across the lipid layer through which water can move freely; if so, the total area of these channels must be 100,000 times smaller than the total area of the cell surface). In each case of passive transport, the molecules diffuse from a space where their concentration is high to a space where their concentration is low, until it is equal on both sides of the membrane.

With sodium and potassium, however, some mechanism actually moves ions in and out of the cell against the concentration "gradients." This active transport, which maintains a high concentration of potassium and a low concentration of sodium within the cell, despite opposite conditions on the other side of the membrane, requires the expenditure of energy. And the energy comes from the cell's universal fuel, ATP. The plasma membrane contains certain enzymes which split the chemical bonds of ATP and then use the energy to carry potassium in and sodium out of the cell. The amount of energy involved is surprisingly high: It has been calculated that up to 18 percent of the total ATP required by the brain, for instance, is expended in such activities.

A third type of traffic across the plasma membrane involves the release or uptake of various substances whose molecules are too big to cross the membrane either by diffusion or through pores. Instead, they travel in membranous packages. In protein-secreting cells, for instance, new proteins are wrapped in pieces of membrane in the Golgi apparatus, forming vesicles; the vesicles then move to the cell's surface, where their membranous envelope fuses with the plasma membrane and their contents are dumped outside the cell.

This is called exocytosis. In a reverse process, endocytosis, proteins or other molecules from outside the cell attach to the cell's plasma membrane, which folds in to envelop them; this forms small vesicles which move into the cytoplasm, carrying drops of extra-cellular medium with the substance to be used inside the cell. In all of these cases, matter is brought into the cell or discharged from the cell without interrupting the continuity of the plasma membrane. Something similar occurs during fertilization, when the sperm's plasma membrane fuses with the egg's plasma membrane in such a way that the sperm nucleus can enter the egg without any exposure to the surrounding medium.

Besides directing traffic in and out of the cell in all these ways, the plasma membrane controls the cell's communications system, receiving signals from the outer environment (including signals from other cells) and sending out messages of its own. Some of these signals are carried by the chemical messengers called hormones, which circulate in the bloodstream with instructions to cells in many parts of the body. As these hormones reach each target cell, they generally do not enter it, but trigger intricate responses. Only a few hormones, such as the steroids (including cortisone and the sex hormones) which are derived from cholesterol and thus easily soluble in fat, can actually pass through the cells' membranes. Most hormones, for instance insulin and some neurotransmitters, act by giving signals to specific receptors on the cell surface; these receptors then activate a second-messenger system which orders appropriate changes within the cell.

Earl Sutherland of Vanderbilt University discovered how this second-messenger system works and won the Nobel Prize for it. He found that adenylate cyclase, an enzyme in the plasma membrane of many cells, can be activated to convert ATP (the universal currency of energy in cells) into cyclic AMP, a nucleotide which is then released into the interior of the cell. Thus, when a hormone (the first messenger) binds to a specific receptor on a cell surface, it triggers the dispatch of cyclic AMP (the second messenger) to the cell's biochemical machinery. Another cyclic nucleotide which acts in similar fashion, cyclic GMP, has been identified and recently scientists discovered yet a third type, cyclic CMP.

It is now clear that these second messengers regulate hundreds of diverse processes in living cells. The surge of energy which we sometimes feel under stress (as a result of the conversion of stored glycogen into glucose); the discharge of secretory products from cells; the activation of dormant genes; and even contact inhibition -- all depend on the action of a second messenger on different components of different cells. The second-messenger system is involved in so many diseases, in fact, that substances which facilitate or inhibit it are expected to prove very useful in medicine in the future.

Hormonal messages travel far and wide in the body via the bloodstream; they can be recognized by very distant organs, but they may take hours, days or weeks to produce their complex effects. In order to stay alive, animals must have a much quicker means of response, with which they can react to events within seconds or milliseconds. This is where the nervous system comes in. When you touch a hot stove, look or listen, the information is carried almost instantaneously by neurons (nerve cells) which, unlike most other cells, produce electrical signals in response to physical or chemical stimuli. The neurons' plasma membranes play a key role in this transmission.

One common sequence is as follows: An electrical signal travels down the surface of a neuron's thin axon (a long fibre); this induces vesicles which contain units of neurotransmitter to fuse with the plasma membrane at the terminal of this axon and to release their contents into the synaptic cleft (the gap between this neuron and the next cell). If enough of the newly released transmitter reaches the plasma membrane of the next neuron, it changes the membrane's electrical potential, producing a new nerve impulse which then repeats the whole process. If, on the other hand, enough transmitter reaches the plasma membrane of a muscle cell at a special region called the neuromuscular junction, it stimulates the muscle to contract.

NEW TOOLS WHICH OFFER A NEW VIEW OF MEMBRANES

As important and powerful as the plasma membrane is, it makes up only a fraction of the total amount of membrane in a cell. Most animal cells, especially those that have specialized functions, are so packed with membranes that there may be 50 to 100 times as much membrane inside them as on their surface. (In plant cells the proportion of inner membrane is even greater, going up to 150 times the amount of membrane on their surface). Yet the plasma membrane is best known to scientists. The main problem in studying membranes has been the difficulty of obtaining pure specimens. Until recently, this was possible only with the plasma membrane of the mammalian red blood cell, which has no nucleus.

When biochemists grind up cells and centrifuge their components, only fragments of the cells' membranes can be recovered -- usually in the form of small, closed vesicles or ragged pieces of unidentified origin. These can be sorted out and identified, so that by now the membranes of practically every organelle can be isolated, but they still vary in purity. Such brutal treatment is not necessary with the mammalian red blood cell, however. Since it contains mostly hemoglobin and has neither nucleus nor mitochondria, nor any internal membranes, it can be emptied with relative ease: It is simply induced to swell up until temporary, large pores open in its plasma membrane, letting the fluid and hemoglobin seep out. The plasma membrane is then allowed to shrink back to its normal size and the pores close, leaving researchers with an intact outer membrane or "ghost." Although other kinds of membranes can now be isolated by various means, the mammalian red blood cell "ghost" remains the favourite subject of many scientists who work on membranes.

Studying these membranes under an electron microscope proved frustrating, however. Since electrons can pass only through ultra-thin sections, the cell was usually sliced finely with a straight knife to prepare samples for the microscope. Yet cell membranes are far from straight and it was practically impossible to get a fullface view of a membrane in this fashion. Nor could one learn much about the membrane's inner layer.

In the early 1960's a Swiss scientist, Hans Moor, developed a method which offered an entirely new perspective on cells. Instead of chemically fixing and staining the cells, he froze them rapidly, following a technique which had previously been used in California with viruses. Then, instead of cutting the cell tissue into smooth sections, he fractured it along natural lines of weakness and bombarded the exposed surfaces with vaporized carbon and platinum, which condensed and hardened on these frozen surfaces. After he allowed the specimens to thaw out, there remained a very thin and highly detailed metal replica of the fractured plane -- a replica thin enough to be viewed under an electron microscope.

Moor naturally assumed that he was looking at replicas of the cells' outer surfaces. But in fact, as Daniel Branton, who was then at the University of California, Berkeley, suggested in 1963, an extraordinary thing was happening: the membranes were being cleaved down the middle in two separate layers. "This seemed a very outrageous idea at the time and no one believed it," recalls Branton, who is now at Harvard University. Yet there was a logical explanation for it: The two lipid layers in the centre of the membrane were only weakly held together by hydrophobic interactions and when the water froze, these interactions became irrelevant, so the membrane fell apart. By 1967 most researchers were convinced. As they began using the freeze-fracture technique to explore the interior of membranes, they discovered little balls -- protein particles -- some of which were invisible from the outer surface. Furthermore, they found that the outer half of the membrane sometimes differed quite radically from the inner, cytoplasmic half, both in protein structure and in lipids. This enabled them to zero in, for the first time, on specific proteins and lipids in the centre of the membrane.

While the freeze-fracture technique opened up the interior of the membrane for study, the true surface of cell membranes became visible through a special kind of electron microscope, the scanning electron microscope (SEM), which provides three-dimensional images. Unlike the transmission electron microscope, whose beams pass through the specimen, the SEM depends on low-energy electron beams which scan the specimen's surface, exciting secondary electrons whose pattern produces the image. Until recently, the SEM could not resolve more than 150 angstroms. "The pictures were gorgeous, but there was very little information coming out of them," recalls Palade.

However, a new instrument developed in Japan is presently making it possible to see down to 30 angstroms. "So now we can start seeing the molecular details of the cell surfaces," he says. "We know that receptors in the plasma membrane must protrude from the surface of cells, but we don't know the size of these protrusions. We want to be able to obtain information about them, not only by biochemical techniques. We want to know the frequency and density of these molecules, their pattern of distribution on the cell surface, whether this distribution can be disturbed and whether any disease states are connected with defects in it. This is the excitment about the high-resolution SEM: It brings you down to a level at which you can really look at molecules!"

To make full use of this instrument's power, the procedures for specimen preparation must also be refined by a factor of five. This requires new methods of "fixing" the specimens with chemicals and also new ways of spraying their surface with metal vapour (to make them give off secondary electrons).

One advantage of working in an interdisciplinary centre such as the Yale Centre for Research on Cellular Membranes is that it is large enough to maintain experts who can devise such techniques. The Centre consists of seven different research groups, each with its specific programme and each depending, at least in part, on advanced technology in two areas: protein chemistry and the SEM. It includes some twenty-five cell biologists, biochemists, pathologists and biophysicists. "None of us uses the SEM or protein chemistry facilities full time," Palade explains, "and none of us could run these facilities with an adequate amount of attention and competence by ourselves. Here we can all use them profitably."

The Center's protein chemistry lab routinely analyzes extremely small samples of protein extracted from cell membranes. It boasts a very effective amino-acid analyzer, controlled by a computer, which uses highpressure liquid chromatography to identify the different amino acids in a protein quickly and accurately. It also has a sophisticated machine which can determine the sequence of these amino acids from a sample of 50 micrograms or even less. Soon the Centre hopes to have an additional instrument to assess the radioactivity in labelled amino-acid residues -- an important tool with which to study the origins and fate of membrane proteins.

Palade's own group of 10 scientists concentrates on unraveling the "functional interaction" between cell membranes and on the membranes' genesis. "The membranes that form compartments within cells are usually separate from one another, but some of them fuse, establishing continuity,"

Palade says. "This is important for secretory processes, when a cell product moves from one compartment to another and then is discharged. The membranes must know the partner with which to fuse; how do they communicate? Every time a cell divides, too, the plasma membrane fuses to split the cell and then the two daughter cells separate.

We are trying to understand this fusion-fission process. At present we only understand the recognition of a hormone or a neurotransmitter by a receptor on the plasma membrane. But we believe the same kind of thing happens inside the cell when membranes fuse, except that both molecules -the key and the lock -- are bound to membranes. The basic principle is the same: recognition by complementarity."

Where do all these membranes come from? As Palade points out, "The cell has discovered a way to increase its membranes without disturbing their function. It appears to use always its pre-existing membranes, which behave like a fine film of olive oil and puts new lipid and protein molecules into it, so that the film expands without ever breaking its continuity. Cells never begin making a membrane from scratch. When they divide, they give their membranes in equal parts to their daughter cells. In fact, we inherit our membranes from our mothers.

"All these membranes have their components replaced continually," he adds. "So there is a continuity of pattern, though not of substance. What we want to find out is where the components -- the proteins and the lipids -- are produced and how they are assembled. We'd particularly like to know where the glycoproteins (sugarbearing proteins), some of which seem to be the receptors of the plasma membrane, are synthesized.

"It's partly a matter of scientific curiosity, but there are also practical reasons: When cells have defective receptors, the actual defect may be inside the cell, rather than on the surface. Maybe the components are not produced in sufficient quantity. Maybe they are produced but are not transported or assembled properly. This may be relevant to cancer, as well as to questions of growth and of aging."

MEMBRANE PROTEINS

One of the Yale Center's most exciting findings so far has concerned the role of "glycophorin," a sugar-bearing protein that spans the entire thickness of the red blood cell membrane. Named by Vincent Marchesi, who succeeded in detaching it, intact, from membrane ghosts, glycophorin has some carbohydrate chains which protrude from the membrane surface like antennae and serve important functions: They act as the cell's recognition sites for blood groups and for certain viruses. In addition, they are involved in maintaining the negative charge on the cell's surface which prevents it from clumping with other red blood cells.

Glycophorin is now the best-known of the glycoproteins that extend through the cell membrane and the first to have its amino acids completely analyzed and sequenced. Immense quantities of blood were needed to achieve this knowledge. (Twelve gallons of human blood can provide only 125 grams of membrane ghosts, which produce only 3 grams of pure protein for analysis). Just recently, Marchesi's research group made the further discovery that under certain conditions glycophorin exists in twin form, with a narrow space to separate the twin shapes, creating an open channel all the way through the membrane. Marchesi believes that these open channels are the long-sought "pores" through which materials can flow rapidly across a membrane. He also believes that glycophorin has brought him to the verge of answering several important and interrelated questions about health and disease, all of which involve the mechanism of cellcell recognition.

For example, there is the mystery of blood platelets - billions of little bodies that float about in blood and pile up at cuts in the blood vessels to prevent the vessels from leaking. Normally these platelets do not stick to one another, but somehow, when they sense an injury to the blood vessel, their surface membranes change and they begin to clump. The mechanism involved here is very similar to that of normal cell-cell recognition, Marchesi points out. Yet in certain cases it may aggravate arteriosclerosis and other forms of cardiovascular disease.

Then there is the lure of understanding the real function of carcinoembryonic antigen (CeA), a protein found on the plasma membrane of some human tumors as well as on normal embyronic tissue. Surprisingly, CeA is an almost exact duplicate of glycophorin, differing only subtly in composition and immunological reactivity. Yet it can be used to measure the progress of certain kinds of cancer therapy: After an operation on colon cancer, for example, the level of CeA in the blood drops to almost zero, going up again if new tumors begin to grow. Researchers would like to develop other CeA-based tests which might reveal the presence of cancer very early, when it is easiest and safest to treat. This might be done through annual blood tests. However, they are hampered by "the lack of a deep enough understanding of what's going on in normal membranes," says Marchesi.

Current methods of analyzing membrane proteins are "much too simple, because what we can do is only twodimensional," Marchesi explains. "We can see only the gross differences between molecules. But there are very subtle ways in which these little chains of peptides can fold over each other. That's the exciting part! We need to develop methods for studying these membrane proteins in three dimensions. There's an infinite amount of complex communication going on between cells, yet probably only a few molecules mediate it, like a translation system or a code. So that's the trick -- to find the code. As you know, the DNA code is only four letters, in a linear arrangement. This code, we think, is hooked up in the 3-D arrangement of the molecules. It's probably going to be a lot more subtle than the code for DNA."

New techniques are being developed towards this goal by Lubert Stryer, formerly of the Yale centre and now at Stanford Medical School. The only proteins whose 3-D structure has been worked out by scientists so far are some 20 water-soluble proteins -- for example, hemoglobin, Stryer says. Analyzing the 3-D structure of insoluble proteins, such as those in membranes, will pose a far greater challenge, he notes, since at present these proteins cannot be crystallized. Researchers in Stryer's lab are starting out with various techniques taken from physics. They are doing optical rotation experiments to see whether glycophorin has any regular structures such as a helix.

They are using a Raman spectrometer, shining laser light of a known wave length onto a glycophorin sample and studying the pattern of light that it scatters as a clue to its molecular structure. They are inserting fluorescein-labelled glycophorin into an artificial membrane, to find out whether it acts as a channel for sodium and potassium and whether altering the part of the molecule that protrudes on one side of the membrane will affect the part on the other side. They are also collaborating with the Brookhaven Laboratories in neutron studies of the molecular structure of rhoclopsin, a membrane protein that is involved in vision.

"All this technology is very, very recent," says Stryer, "but it may be extremely powerful. We are building up an image of how rhodopsin sits in this membrane -- it may actually go through the membrane. If rhodopsin turns out to be a gate that is opened by light, other gates in membranes may be opened by hormones in a similar way. That's our hope."

THE PROMISE OF NEW THERAPIES

Throughout the world, researchers who work on cell membranes realise that they have only just begun to explore a field in which anything seems possible. They share a great excitement about it.

"Can we expect to do 'membrane engineering' in the future?" S.J. Singer asked a scientific gathering recently. And he answered with a resounding, "Yes!" For example, it may be possible to change the mobility of membrane components, he says. This might have extraordinary results. Normal cells have some kind of matrix underlying their membranes which restricts the mobility of certain membrane components, he points out. Yet when cells are malignantly transformed, their membrane components can suddenly move about more easily. This on-off control of the components' mobility may well be the mechanism of growth, Singer believes; the oily film of lipids in the membrane may need to become more fluid to allow for the insertion of new molecules as the membrane grows. The same on-off control might also be a key to cancer. Scientists might then restore malignant cells to normality by providing anchoring points for the components of their plasma membranes.

Physicians already do modify the plasma membrane of their patients' cells when they use various general anesthetics, which dissolve in membrane lipids and change their properties. In the future, it is almost certain that drugs will be tailor-made to act more selectively either on proteins or on lipids in the plasma membranes of specific kinds of cells.

Some of these drugs may themselves be enclosed in artificial membranes, forming "liposomes." In this way they could be engineered for release only when or where needed, by means of specific receptor molecules which would be inserted in the liposome membrane. "For example," suggests Cambridge University's Alec Bangham, "liposomes containing a tranquilizer could be equipped with molecules that only respond to supernormal levels of adrenalin." Similarly, liposomes loaded with antitumor agents could be keyed to the properties of specific tumor cells so that the drug would be released only when the liposome came in actual contact with the malignant cell. "By this method," Bangham explains, "It would be possible to employ drugs far too toxic for administration through the general circulation." Experiments of this sort are already under way in Christian de Duve's lab and in other places.

Other researchers have been working on ways to correct the intricate feedback system through which cells control the production of such vital chemicals as cholesterol, in cases when the system fails. As Joseph Goldstein and Michael Brown of the University of Texas discovered recently, each person's production of cholesterol is controlled by the activity of specific receptors for LDL (low-density lipoproteins, the particles which carry cholesterol in the blood) in his or her cells' plasma membranes -- and these receptors in turn increase or decrease according to the amount of cholesterol in the cell. Mammalian cells can both produce their own cholesterol and take up cholesterol from the blood. When a normal cell needs more cholesterol than it has produced, it synthesizes more LDL receptors. These transmit LDL from the blood to the cytoplasm, where lysosomes break it down, releasing cholesterol. As soon as the cholesterol in the cytoplasm reaches a certain concentration, however, it shuts off the enzyme that produces cholesterol inside the cell -- and also stops the synthesis of additional LDL receptors.

The system works pretty much like a thermostat to ensure that the cell always has enough cholesterol for life and growth, without accumulating too much of it. Because of a defective gene, however, some people do not have a sufficient number of LDL receptors in their cells. This is what happens in familial hypercholesterolemia, where the lack of LDL receptors breaks the normal chain of control and leads to overproduction of cholesterol. Among people who suffer heart attacks before the age of 60, one out of 20 have been found to carry this defective gene (which actually occurs in about one out of 500 persons in the general population). Now that scientists understand the feedback system involved, they realise why existing anticholesterol drugs are so often ineffective: lowering the amount of cholesterol in the blood may actually lead the flawed cells to produce even more of it, to make up for the loss. By contrast, new drugs which are now being tested would specifically reduce the cell's ability to make cholesterol, rather than remove more of the cholesterol that is already made.

Many forms of mental illness may also turn out to be related to defects in specific proteins that act as receptors for hormones and neurotransmitters on the plasma membrane. If so, it may be possible to develop far more effective drugs which would take account of the cell's feedback systems. Every new bit of information about the inner workings of the cell leads towards the development of better weapons against the diseases that still plague us. As researchers learn more about membranes, for example, they are indirectly contributing to the prevention or cure of such seemingly unrelated ailments as diabetes, heart disease, various genetic defects, cancer and even kidney disease, for the kidney is composed largely of membranes. New vaccines and new immunologic weapons also await a better understanding of the cell and its organelles.

While today's physicians view diseases in terms of organs, such as liver diseases, bone diseases or heart diseases, the physicians of the next generation will see these ailments as diseases of cell membranes and organelles -for example, diseases of the endoplasmic reticulum, lysosomal diseases, or mitochondrial diseases. Thus, heart disease will be examined not simply in terms of which chamber and portion of the heart is involved, as determined by the EKG, but in terms of the specific lesion within the heart muscle cells that leads to malfunction and cell death.

This has been a period of "spectacular growth" in cell biology, the President's Biomedical Research Panel declared recently. "Basic biological and biomedical sciences have progressed faster over the last 25 years than at any time during their long history... For the first time in the history of the life sciences, the essential function of cells can be understood in terms of defined chemical reactions." Yet despite all this progress, most of medicine is still far from being a true science.

The reason for this gap is that, until now, scientists could not even study the mechanisms which regulate such basic processes as cell growth, division, or mobility; they simply did not have enough information about the cell's components. They are barely starting to understand how cells are controlled. But now the stage is set. A far more powerful type of research on the cellular and molecular basis of disease can begin. And as it builds up a clearer picture of normal and abnormal cell function, it will create a truly scientific medicine for the 21st century.

Chapter 5

The Golgi Apparatus

The Golgi apparatus was a structure first described in 1898 by the Italian neurologist Golgi, in the nerve cells of the Barn Owl. Since the discovery of this organelle it has been the subject of very great controversy and in the last 50 years there have been at least 2000 papers written about it. The standard technique for differentiating the Golgi apparatus has been by the use of osmium or silver salts after appropriate fixation of the cell and the most characteristic form for the organelle which has been demonstrated is that of a network.

In the case of the nerve cell, this network extends pretty well through most of the cytoplàsm of the cell, in other cells it is a small compact area situated to one side and closely applied to the nucleus. The conception that the Golgi apparatus existed as an osmiophilic or argentophilic network was challenged by a number of workers.

Parat and his school believed that the Golgi apparatus really consisted of a series of vacuoles which stained with neutral red and an extension of this hypothesis, e.g., by John R. Baker of Oxford, conceived of the network actually being produced by the deposition of metals, osmium or silver, on the periphery of the vacuoles so that eventually a networklike structure was built up.

Very complex functions were deduced for the Golgi apparatus and it was believed to play a part in the formation of the acrosome of the sperm and to be concerned with yolk reproduction of the egg and secretion products in other cells, particularly those of the glandular cells.

Dr. Pollister in the *International Review of Cytology* has

discussed the structure of the Golgi apparatus and pointed out that many authors have described the apparatus as being lamellar in shape. He pictures it as being composed of irregularly circular, flattened, lamellae which surround one end of the nucleus. In various studies Pollister attempted to measure the thickness of these lamellae and apparently they were in some cases thinner than lu and in others around 0.25 or 0.20u. The apparatus is capable of considerable distortion and can be seen, in fact, distorted in the contracting smooth muscle fibre but appears to return to its normal shape once the force which is altering its shape is removed.

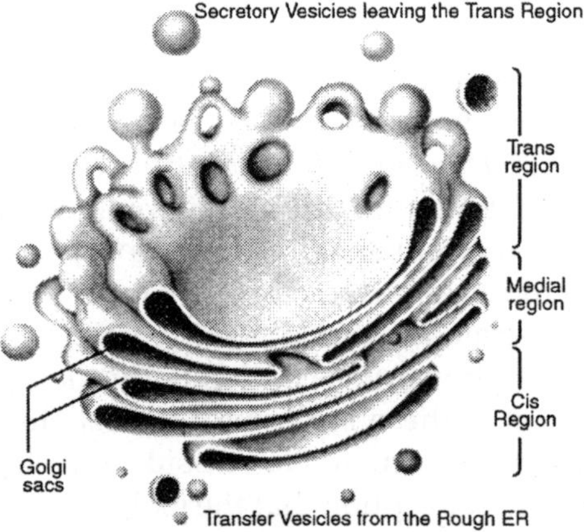

Fig. The Golgi Apparatus

Possibly one of the most interesting developments of the study of the Golgi apparatus was its isolation from the epithelium of the epididymis of mouse and rat by Dalton and Felix in 1954. This was obtained by differential centrifugation from homogenates. It is also of interest that the Golgi apparatus appears to resist a good deal of damage to the cell.

As long ago as 1925, Avel showed that, if living cells were ruptured by osmotic means, the Golgi apparatus was often one of the last parts of the cell to undergo disintegration. The Golgi apparatus was extremely difficult to see in the living cell under the standard techniques of optical microscopy even with the phase contrast microscope; (however, with better techniques it was visible in germ cells). The whole controversy about the

Golgi apparatus appears to have been now reconciled by the studies with the electron microscope. The work responsible for this was done by Felix and Dalton as described in a series of papers published between 1953 and 1957 and also by Sjöstrand. They showed that the Golgi apparatus consisted of a series of pairs of membranes adjacent to the nucleus which contained dilatations giving the appearance of vacuoles. This structure thus explains many of the controversial results of the early workers. Sjöstrand, who contributed to the elucidation of this problem with his colleague Hansen, has stated that the membranes which constitute the Golgi apparatus form a system (about 60 A thick). These membranes are arranged in pairs. Along their edges there may be points of fusion followed by dilated areas (the vacuoles).

The width of the space between the pairs of membranes (60 A) is, apart from the presence of vacuoles, usually fairly constant. These pairs of membranes seem to be embedded in ground substance which is fairly homogeneous and has little structure, but in some cases it seems to contain fine granules or a fine reticulum.

One of the characteristics of the Colgi apparatus which might otherwise have led it to be confused with the E.R. membranes is that the space between the latter, as previously described, measures 150 A. Furthermore ribonucleoprotein granules are attached to the outside of most of the E.R. membranes, but the membranes which constitute the Golgi apparatus are quite smooth and have no granules associated with them although there are reports of relatively large granules (400 A across) being situated on or near the pairs of membranes in some cases.

It is interesting that this characteristic structure which has been found for the Golgi apparatus is pretty well uniform in the cells of most species of animals and in cells belonging to a wide variety of organs; for instance, gland cells, nerve cells, mucle cells and most of the other cells in the body show the same sort of general Golgi picture.

The Golgi apparatus has been characterized throughout the literature as being that part of the cell in which the products of secretion are first recognizable microscopically and Hirsch has even described in this living pancreatic cell small granules which were attached to the surface of the mitochondria becoming detached and moving through the cell cytoplasm toward the Golgi material.

The amount of information on the chemical nature of the Golgi apparatus is very large and very confused. We have just noted that the Golgi apparatus is composed of paired membranes which are dilated in parts to give the appearance of spheres, but this description of the Golgi apparatus does not satisfy all the various descriptions of the forms which it takes in different cells when viewed by the light microscope. The relationship of this suggested ultrastructure to, for instance, the production of oögenesis and spermatogenesis has yet to be worked out.

It is of interest here to note that Dr. H. B. Tewari has recently demonstrated that in the langur monkey, *Semnopithecus*, Golgi bodies pass from the follicular cells of the ovary into the developing cocyte where they appear to play a part in the elaboration of yolk. The passing of these bodies from one cell to another is the really fascinating part of this observation.

The electron microscope will probably, in due course, give us some information on the nature of Hirsch's so-called presubstances and what relationship they bear to a fully devoloped Golgi system. The significance of the difference in the physical nature of the apparatus in different cells will need to be explained, too. For instance, in most cells the specific gravity of the Golgi apparatus is less than that of the other cellular constituents but in uterine gland cells this is not so. In the cells of these organs it seems, in fact, to be a relatively rigid structure.

Pollister has suggested that the Golgi apparatus could not be a fluid or even highly plastic solid but that it has fundamentally a platelike form with enough elasticity to bend under pressure and straighten out again when the pressure was removed. Simpson, on the other hand, has claimed that it was always in a highly fluid condition. It is obvious that a good deal of the Golgi problem has to be reexamined and many light microscope studies remade in view of the findings of the electron microscopist.

The ability of the Golgi material to reduce osmium tetroxide which it does very rapidly is an indication that it may contain unsaturated lipid but then, of course, any reducing substance will reduce osmium tetroxide so this in itself is not a clearcut histochemical test. However, the Golgi material has been stained but rarely with fat dyes, nevertheless the fact that it is soluble in the usual fat solvent suggests that it may contain fatty or lipoidal material. It has been claimed that the Golgi material breaks up after narcosis with chloroform. Monet has shown, that if the germ cells of Helix are treated with sodium bicarbonate, myelin figures are formed from the representatives of the Golgi material (dictyosomes).

On the other hand, it has been claimed by Thomas that the dictyosomes of Helix are not Golgi material at all but are formed by the overimpregnation of mitochondria; this is the type of confusion that bedevils the subject of the Golgi apparatus. Ciaccio stained the Golgi apparatus and spermatids with his lipoid technique and Boyle claims that part of the Golgi apparatus of the neurons of Helix are stained by Sudan IV which is a fat stain. Baker carried out a series of detailed histochemical tests on the Golgi material and he found that the apparatus appeared to contain lecithin, cephalin or sphingomyelin and that Windaus' test for cholesterol and Schultz' test for cholesterol were negative. From the foregoing it seems reasonable to suggest that lipoidal material is present in the Golgi apparatus. As long ago as 1925, Nath suggested that the Golgi material contained protein.

The same suggestion had been made by Bowen. Gatenby has expressed the opinion that the Golgi material is a combination of protein and lipoid. In fact some authors have found it possible to demonstrate the Golgi apparatus by fat dyes following a prior treatment of the tissue with proteolytic enzymes (pepsin and trypsin). They suggest that these removed the protein which was masking the lipoprotein complex and preventing the lipoid from reacting with the fat dye. Baker using a variety of histochemical tests found that the Golgi apparatus did not contain arginine or glutathione but that it gave a positive reaction with Millon's reagent, a positive xanthoproteic test and also a reaction for tryptophan. However, he found that the Golgi material was not coloured more intensely than the cytoplasm so there was not a particular concentration of the amino acids which give these reactions, in the Golgi apparatus.

The localization of vitamin C in the Golgi apparatus has a considerable literature and has been the subject of considerable controversy; for the latest discussion on this subject the reader is referred to a publication by the present author in *Protoplasmatologia "Vitamin C in the Animal Cell"* and a comparable article in the same volume by Plaut on "Vitamin C in the Plant Cell". Evidence for the occurrence of vitamin C in the Golgi region has also been presented in the chapter on *"Mitochondria and the Golgi Complex"* in *"Cytology and Cell Physiology"*.

There appears to be a considerable identity in several types of cells between the Golgi preparations and preparations which are demonstrated by the application of the vitamin C reagent (acid silver nitrate). For example, in the neuron of the developing chick the Golgi preparation resembles very closely the result which one gets from application of vitamin C reagent and in the chick embryo liver a similar correlation can be seen. Diuresis causes changes in position of the Golgi material in rat kidney cells and this is comparable to the distribution of the vitamin C reaction. There is also a similarity of distribution of Golgi material and vitamin C in fibroblast cells and in Goblet cells in the rat colon.

Furthermore the distribution of the vitamin C reaction in the ultracentrifuged adrenal, cortical and medullary cells is identical with the position that is obtained with the Golgi stain. However, as has been mentioned before, the vitamin C reagent is extremely destructive of the cytoplasm of cells and it is very difficult when looking at the effect under the electron microscope to be dogmatic about localization of this material in the Golgi apparatus. Until further evidence is obtained, it is necessary to be conservative about the interpretation of vitamin C reactions in cells. Some enzymes have been demonstrated to be present in the Golgi apparatus, for instance, it was first shown by the present author in 1943 that the columnar epithelial cells of the guinea pig jejunum contained alkaline phosphatase in the Golgi region. The localization of this enzyme in the same region in the cells of the mantle edge which secretes the shell of the mollusk *Mytilus* was also figured. Various other authors have subsequently recorded the presence of this enzyme in the Golgi region. Deane and Dempsey, for instance, found that alkaline phosphatase was distributed in the Golgi region of the duodenal epithelial cells as granules or as a continuous reticulum. Activity was most intense in the Golgi region at the bases of the villi. Enzyme granules were found in the Golgi region in the cells of the kidney tubules, bile capillaries and uterine epithelial cells of various mammals and acid phosphatase has also been demonstrated in the Golgi region of the duodenal cells and in the uterine epithelial cells of pregnant cats and sows.

Acid phosphatase has also been found in the Golgi region of the ventral lobe of the prostate by Brandes and Bourne and details of the influence of the male sex hormone in maintaining its presence in this organelle has already been described. It is of interest that the type of castration change described for the prostate could not be detected by biochemical studies of the cell, so that a biochemical investigation of the effects of castration or male sex hormones on acid phosphatase in the ventral lobe of the prostate would probably have shown no significant change in the level of acid phosphatase in the cells. Yet when this is examined by histochemical methods it can be seen that the most fundamental changes have in fact taken place in the cell and a basic change in the locus of activity of an important enzyme has occurred.

Deane and Dempsey have also demonstrated that adenylic acid phosphatase (presumably 5-nucleotidase) has a similar distribution in the Golgi apparatus of liver cells as glycerophosphatase but it appears at a different pH. These authors suggest that all cells may have significant phosphatase activity in the Golgi zone at some pH and with some substrate. As far as the function of the phosphatase in the Golgi apparatus is concerned nothing certain is known.

Emmel has pointed out that the enzyme is present in the lumen of the intestine, kidney tubules, uterus and bile cavities and suggests that it is being synthesized and excreted by the Golgi complex. This is of course a possibility. It may also be concerned, particularly in the absorptive cells of the gut, with the phosphorylation and dephosphorylation process concerned with the passage of some molecules through the cell membrane.

In the last few years the Golgi material which Dalton and Felix have been able to isolate from homogenates of epididymides has provided more specific information concerning its composition—at least in this organ. They found the isolated apparatus to be refringent and part of it to be extractable with 70% alcohol.

Another part, however, was insoluble but this part was stainable with Sudan black, indicating some lipid content; presumably it was a lipoprotein. Schneider and Kuff found with identical material that the pentose nucleic acid (RNA), the phospholipid and the phosphatase concentration in the isolated Golgi fraction was greater than that in the whole tissue. Ascorbic acid, DNA, cytochrome oxidase and DNAase were absent from the Golgi fraction, but the isolated Golgi material also gave a strong periodic-acid Schiff reaction which seems to have been due to a lipid component. A similar reaction has also been reported in intact cells from Leblond's laboratory.

In addition to the phosphatase which the Golgi apparatus of most cells has been shown to contain, other enzymes may occur. The present author's studies have demonstrated that oxidative enzymes are often present in the Golgi apparatus of Purkinje and other brain cells. It looks as though the composition of the Golgi apparatus from an enzymic point of view may differ from cell to cell and possibly from time to time, but further studies of this are required.

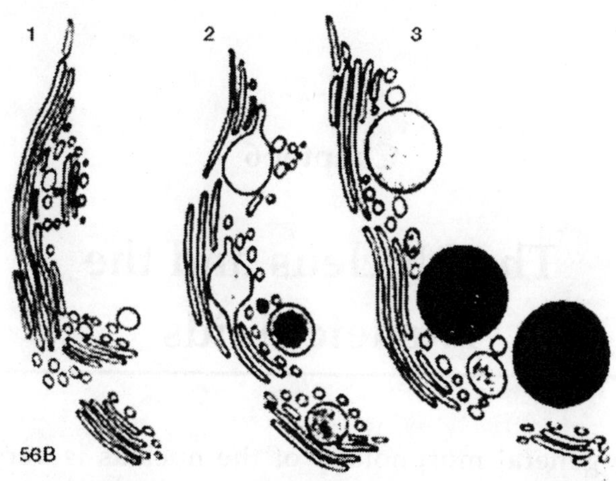

Fig. Production of secretion droplets in Golgi lamellae.

1 System is resting.

2 The system is building Golgi vacuoles and intermediate bodies.

3 Zymogen granules with membranes produced by Golgi lamellae.

Chapter 6

The Nucleus and the Nucleic Acids

The general morphology of the nucleus is very well known and need not be considered in great detail here. Under the light microscope it appears to have little structure apart from the existence of refringent nucleoli. In certain kinds of cells, for example, eggs, the existence of nuclear membrane can be detected with the light microscope.

Under phase-contrast microscopy more detailed structure can be seen in the nucleus itself and most of this represents the chromatin which with appropriate fixing and staining appears as a network. The spaces between the network contain what might be described as a "sap" which varies in amount in different types of cells. The shape of the nucleus is usually spherical but may be altered in pathological conditions and in some organs of senescent animals distorted nuclei can be seen. In old cells the nucleus may become pyknotic and stain excessively with basic dyes and under certain physiological conditions the nuclear shape may change—a classic example of this is the shape of the nuclei of the silk gland cells of the silkworm.

Here the tortuous shape of the nucleus, involving as it does a tremendous increase in the area of the membrane, must indicate a considerable degree of nucleocytoplasmic interplay. This subject will be referred to again in much greater detail. The shape of the nucleus may vary from a rounded spherical to even a branched shape as seen in the silk gland cells of the silkworm or it may even be segmented or have any kind of irregular shape as, for example, in the polymorphonuclear leucocytes of the blood.

The nucleus is known to contain two types of nucleic acid, deoxyribonucleic acid and ribonucleic acid. It contains basic proteins and other proteins which include enzymes and, phospholipids, various phosphate compounds and a number of inorganic compounds.

There is now a considerable list of enzymes which have been found in the isolated nucleus. These include those which are related to the glycolytic cycle and to the oxidative cycle (although the latter are present in very much smaller amounts than in the cytoplasm). Enzymes concerned with the nucleotide metabolism are naturally present and enzymes which play a part in protein and fat metabolism also occur. In addition to these biochemical results which have been obtained, studies in the author's laboratory on the distribution of a variety of dephosphorylating enzymes which hydrolyze uridine, inosine, cytidine and guanosine triphosphates and others which dephosphorylate DPN, TPN and a wide range of other phosphate esters are present in the nuclei of most cells.

Generally speaking these results indicate, as one might expect, the presence of most of the enzymes concerned with nucleotide metabolism and with many of those concerned in glycolysis. However there is complete absence of a number of enzymes vital to the functioning of the Krebs cycle, but a surprising concentration of some others such as aconitase. The relatively large amount of cytochrome c is of interest.

In the absence of a Krebs cycle there is some doubt as to whether the nucleus and in particular the nucleolus is able to synthesize ATP, although the nucleus contains plenty of ATPase. If the nucleolus synthesizes protein (evidence for this will be presented shortly) it needs ATP. Where does it get it from? The mitochondria have all the equipment for the production of ATP and it has been noted that in tissue culture cells the nucleolus frequently moves in the nucleus to touch the nuclear membrane. Mitochondria have often been observed to touch the nuclear membrane at the same spot and at the same time as the nucleolus and it has already been suggested that perhaps ATP is passed across the nuclear membrane at this time.

It is of interest that Tewari and the present author have shown that in spinal ganglion cells there appears to be a cycle of activity between the nucleolus and the mitochondria. When the nucleolus moves to the side of the nucleus and touches the nuclear membrane, the mitochondria are clustered around the nucleus. The nucleolus then passes to the centre of the nucleus and the mitochondria disperse through the cytoplasm. At the same time droplets of unidentified material make their appearance in the cytoplasm.

Before 1950, the electron microscope had given us very little help as far as the structure of the nuclear membrane was concerned, but during that year Callan and Tomlin were able to dissect out the large nuclei of the germinal vesicle of the oöcytes of amphibian eggs. These nuclei were ruptured by exerting a slight pressure on them and the broken nuclear membranes so obtained were then examined under the electron microscope. By using this technique they found that the nuclear membrane actually contained pores. Callan and Tomlin then found that the nuclear membrane was composed of two sheets, an outer one in which the pores were present and an inner continuous sheet; these pores were approximately 400 A in diameter.

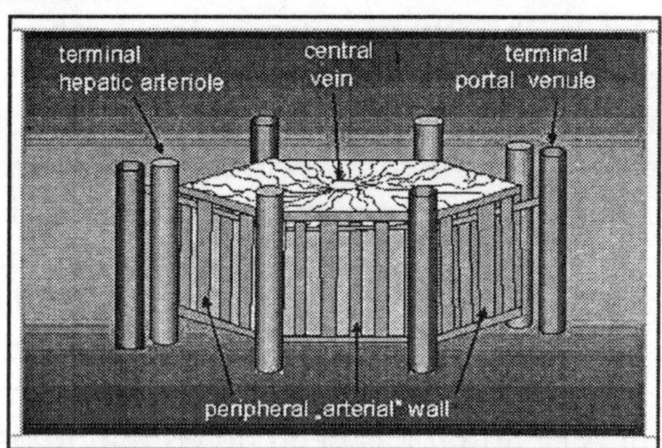

Fig. Electron Micrograph of Rat Liver Cell

Pores of this type have also been found in the oocytes of the starfish and in sea urchin eggs; the nuclear membranes of pancreatic cells have also been said to contain such pores. It has been claimed that neurons possess as many as 10,000 pores in each nucleus. Although there has been some discussion about the reality of the existence of these pores, it is fairly certain now that they do occur more or less universally.

The dimensions of the nuclear membrane are of interest. Under the electron microscope the membrane can be seen to be composed of two osmiophilic layers with a clear layer in-between. Each of the osmiophilic layers has a thickness of 70 to 80 A and the nonosmiophilic space between them ranges from 100 to 150 A. It is of great interest that this is practically identical with the dimensions of the intracytoplasmic membranes (endoplasmic reticulum) and it has, in fact, been suggested by Watson that the nuclear membrane is not a separate membranous structure at all but is simply a part of the endoplasmic reticular membrane. If these can themselves be regarded as extensions of the cytoplasmic membrane then the nucleus is really enveloped in what is a deep and complex fold of the cell membrane.

The specific details of the nuclear membrane vary considerably in different types of cells but, generally speaking, the three-layered structure of the membrane and the presence of pores are characters that can be accepted for most nuclear membranes examined to date. The nucleus of nondividing cells has been referred to as being in a "resting" condition and this is singularly inappropriate in view of the metabolic activity which this organelle must be carrying out almost continuously.

The relationship of the volume of the nucleus to that of the cytoplasm varies considerably in different types of cells and, generally speaking, the volume of cytoplasm increases by comparison with the nuclear volume as the cells age.

Nuclei which are fixed by standard histological and cytological fixatives show a nuclear reticulum on which are scattered basophilic staining substances which have been described classically as chromatin and the parts of the reticulum which do not show basophilic staining have been referred to as "linin." In and around this network it has been assumed that nuclear sap, which has been described as the "karyolymph" or "enchylema," exists.

The other obvious structure in the nucleus which can be seen is the nucleolus. There may be one or more of these and they vary considerably in shape. The nucleolus was first described as long ago as 1781 by Fontana and from then on it seems to have been pretty generally recognized as a typical cell structure. Nucleoli under the light microscope usually have the appearance of rounded homogeneous structures always within the nucleus, but Tewari and the present author have seen them attached to the outside of the nuclear membrane in spinal ganglion cells. There appear to be two structural phases in the nucleolus, one is in the form of a series of highly coiled strands and the other a structureless phase, presumably some sort of a nucleolar sap.

This nucleolar sap is called the "pars amorpha" and the coiled strand is the nucleolonema described by Estable of Monte Video and his co-workers. The nucleolus contains a good deal of ribonucleic acid and there is also some evidence that it contains deoxyribonucleic acid as well. Caspersson has claimed that it is rich in diamino acids and the presence of protein masked phospholipids has also been described; in the spinal ganglion cells the nucleolus stains intensely with mitochondrial techniques, but further details of the chemical nature of the nucleolus will be given later on. The nucleolus varies in shape and size according to the activity of the cell, during the period of anabolism it is said to become hypertrophied and during the catabolic stage it is said to become reduced in size. The nucleolus shows changes with pH, a number of drugs cause it to change in size and it is also affected by ionizing radiations.

Although the nucleolus is an area in the nucleus where there is a great concentration of special substances, the electron microscope up to date has given no evidence that it is surrounded by a limiting membrane, this is not quite what one would expect and so perhaps it is necessary to be careful in discussing nuclear-nucleolar interrelationships until the significance of this fact is assessed. In the interphase nucleus, the nucleolus retains its characteristic spherical shape and, provided the cell is not subjected to any particular type of stress, the size and general appearance of the nucleolus will remain unchanged. During the prophase stage of cell division, the nucleolus demonstrates a relationship to a specific chromosome as it develops.

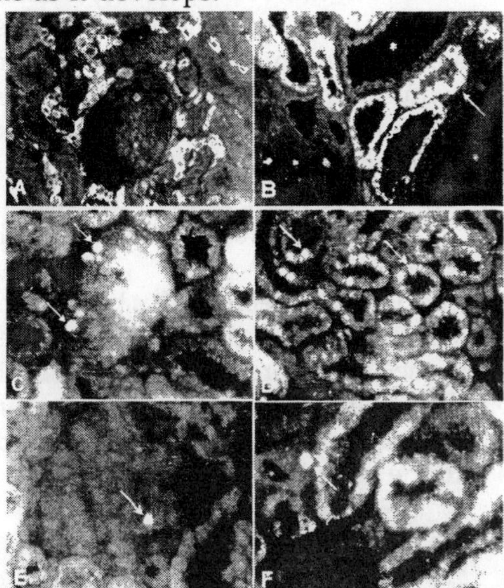

Fig. Electron Micrograph of Tubular Cell from rat Kidney

With the onset of metaphase at a time when the nuclear membrane dissolves away, the nucleolus disappears or ceases to be evident as a formed structure. At telophase, just before the nuclei re-form, small basophilic bodies appear between the chromosomes which form up to produce the nucleolus again. One of the most characteristic inclusions in the nucleolus is the vacuole —a structure which seems very resistant to most of the staining agents. Darker staining areas within the nucleolus are known as nucleolini. Although there is some evidence that the nucleolus is in a semifluid state, it has a considerable density and can be centrifuged to one end of the nucleus and more or less pure nucleolar material can be obtained by differential centrifugation.

The studies with x-ray absorption and ultraviolet absorption suggest that the nucleolus is a semisolid body and that it contains proteins which are in a state of considerable dehydration. It is of interest that in cells which engage in active synthetic activity the nucleoli become hypertrophied and cells which are not actively synthetic have little or no nucleoli. The classic example of this can be found in the case of muscle. Embryonic cells which are forming muscle fibers have very well defined nucleoli, presumably because they are concerned in synthesizing muscle protein whereas the nuclei in muscle fibers of adult animals show very few nucleoli and those which are present are extremely small. Starvation also causes a great reduction in size of nucleoli but they increase in size very substantially on refeeding, suggesting again that protein synthesis is taking place.

In embryonic tissue, nucleoli always become most obvious at the time when the tissue is differentiating, presumably at the time when specially differentiated protein for building up a specific organ is being made. There is some evidence that either whole nucleoli or nucleolar material can be passed from the nucleus into the cytoplasm.

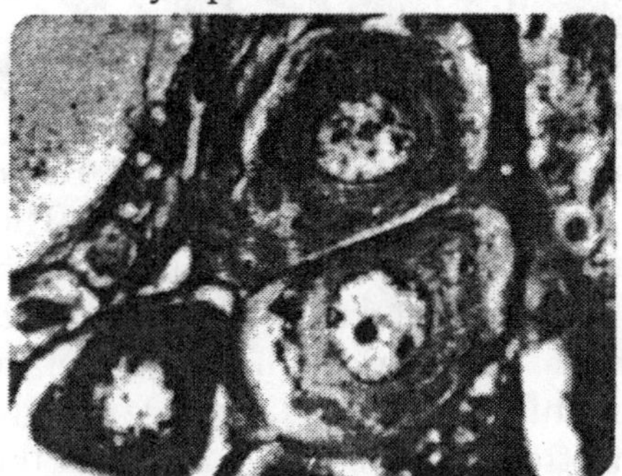

Fig. Spinal Ganglion Cells of Rat

This has been recorded both for fixed and for fresh tissues and Duryee mentions that it takes place in the living oocytes of amphibia. Tewari and the present author have seen it in spinal ganglion cells. It, therefore, appears quite possible that this may occur and will be referred to again later.

The rest of the nucleus is composed of nucleoplasm which does not appear to have a highly organized structure as demonstrated under the electron microscope but rather seems to be built up of irregularly distributed particles of various sizes although their average diameter according to Sjöstrand is 170 A. The range of variation being 150-190 A.

CHROMOSOMES

Chromosomes appear to organize themselves out of the reticulum of the nucleus during the process of mitosis and during the preliminary stage of the prophase they become shorter and detach themselves from any reticular material left in the nucleus. Subsequently, with the disappearance of the nuclear membrane, the nuclear sap diffuses into and becomes mixed with the cytoplasmic materials. At this stage production of a spindle begins and the chromosomes lie free in the cytoplasm and then become organized in relation to the spindle which forms around them. Chromosomes are bodies which have been known for a very long time and even those of the insects have been known since the middle of the 19th century. Insect salivary gland chromosomes are extremely large and show a very complex banding which appears to be related to the location of the genes. These bands appear to contain DNA and can be demonstrated by a number of staining methods.

The persistence of chromosomes in the resting nucleus has been a problem to cytologists for many years and studies with ultraviolet absorption on the nuclei of living cells have demonstrated that at least the DNA of the chromosomes is still present at this stage. If resting nuclei are broken up and ultracentrifuged, fine filaments can be obtained from them and it appears that these are probably greatly extended chromosomes. Certainly there is some evidence that typical chromosome structures are present in them. The bands which are shown to be present on chromosomes are known as euchromatic bands and heterochromatic bands. The former were said to be composed of DNA associated with histones and the heterochromatin appears to contain both DNA and RNA.

Curious types of chromosomes known as lampbrush chromosomes have been described in the eggs of amphibia, fish, reptiles and birds and they, like the other types of chromosomes, are banded but also have a number of loops which project from the surface and extend into the nuclear sap in which the chromosomes lie. These chromosomes are extremely long and very thin and they develop their particular shape just about the time when yolk is being synthesized actively by the egg. At this time, too, the nucleus appears to contain quite a number of nucleoli and it is thought that these are produced by specialized regions in the chromosomes.

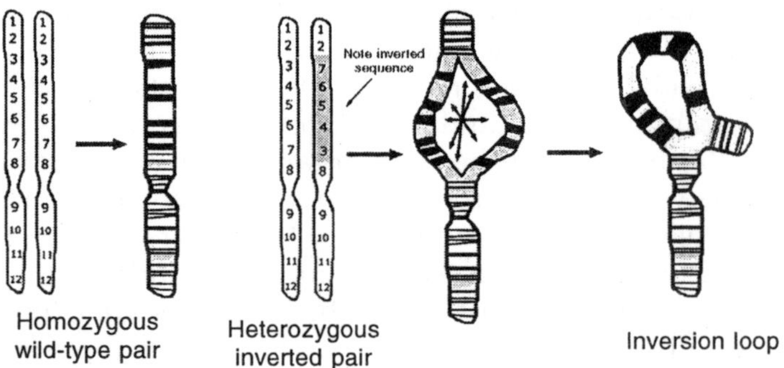

Fig. Portion of a Salivary Gland Chromosome in Chironomus

When oogenesis is complete the chromosomes shrink and become very much thicker and the number of nucleoli becomes reduced. The main axis of the lampbrush chromosome contains a considerable amount of DNA and the loops which project into the cytoplasm seem to be composed largely of RNA. It has been demonstrated by Ficq that precursors of RNA and proteins which have been labeled with radioactive atoms are localized more in the nucleus than in the cytoplasm and that the incorporation of, for example, adenine into RNA seems to be localized almost exclusively in the loops of the lampbrush chromosomes. It seems pretty certain that the loops of these chromosomes represent highly active areas which are specifically concerned with the synthesis of protein.

This discussion of chromosomes leads us on to the concept of the gene. It appears to be generally accepted that genes can be described chemically as nucleoproteins. The difficulty, however, is to find some way of explaining why these various genes are able to exert, either by chemical alterations, or physical structure, their different activities. It has been suggested that the difference in the proteins associated with the DNA in each of the genes is the important factor. At least this was originally thought to be the case, but there is now a good deal of evidence that suggests that the difference is really not one which exists among the proteins.

It is thought that the DNA may vary from the point of view of the sequence of nucleotides in the nucleic acid or in the spatial relationship of the nucleotides to each other; the length of the nucleotide chain which makes up the DNA may also be important. Stern has suggested that one of the analogies concerning the genes was that the relationship of the different organization of DNA to production of different types of genes was "the analogy of the sound track traced in a plastic matrix by a recording stylus, the modulated grooves would be the counterpart of genic modulations engraved on a chemically uniform nuclear protein matrix." According to this point of view the various genes would represent isomers or stereoisomers of each other.

We have mentioned RNA and DNA frequently and we should now consider these substances in a little more detail. They are both nucleic acids and constituents, not only of the nucleus but of the nucleolus as well and RNA is also present in the cytoplasm in association with the endoplasmic reticulum as ribonucleoprotein granules and it also occurs scattered through the rest of the cytoplasm.

The discovery and identification of nucleic acids began with the work of Miescher in the last half of the 19th century and our knowledge of the structure and function of these acids has developed from this point. Nucleic acids are found in all living cells, not only in animals but also in plants and are associated with proteins to form nucleoproteins.

Nucleic acids are actually polymers, being composed of a very large number of nucleotides which represent the monomeric part of the polymer. They all have a basic chemical structure being composed firstly of a base which can be either a purine or a pyrimidine, secondly a sugar and thirdly a phosphate group. There are two types of sugar present in nucleic acids, d-ribose and deoxyribose and nucleic acids have been divided into two main groups according to which type of sugar they possess.

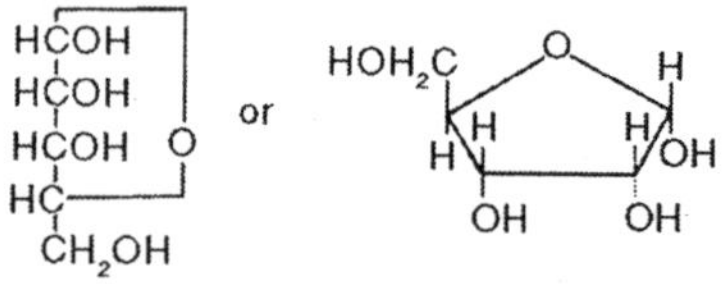

D-Ribose (α-D-ribofuranose)

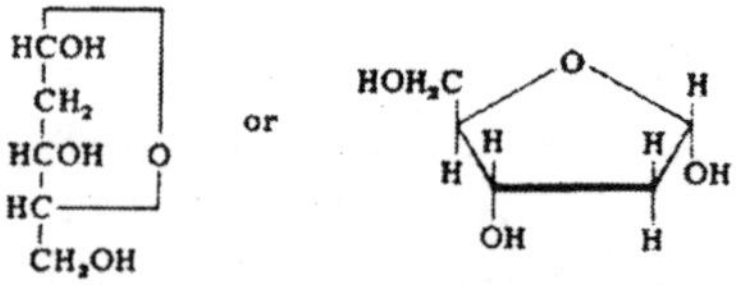

D-2-Deoxyribose (α-D-2-deoxyribofuranose)

Those containing the ribose are known as ribonucleic acid and are represented in short-hand terminology as RNA and those containing the second type of sugar are called desoxyribonucleic acid and are given the symbol DNA. Some people use the term pentose nucleic acid for RNA and give it the symbol PNA and this difference in nomenclature sometimes leads to confusion among the uninitiated in this field. There are 5 bases associated with nucleotides of which two (adenine and guanine) are purines and three (uracil, thymine and cytosine) are pyrimidines.

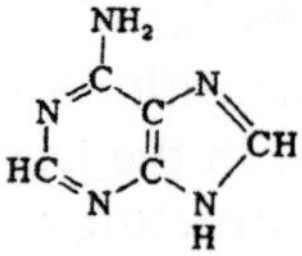

Adenine (6-Aminopurine)

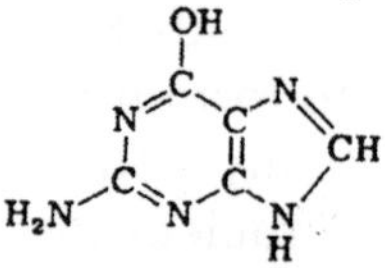

Guanine (2-Amino-6-oxypurine)

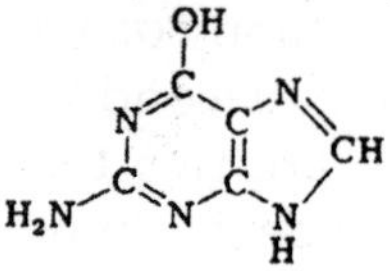

Chromosomes Uracil (2, 4-Dioxypyrimidine)

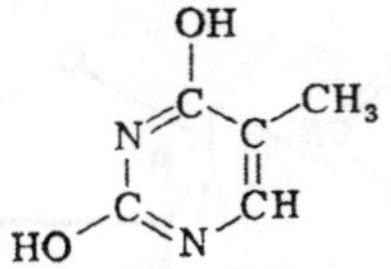

Thymine (5-Methyl-2,4-dioxypyrimidine)

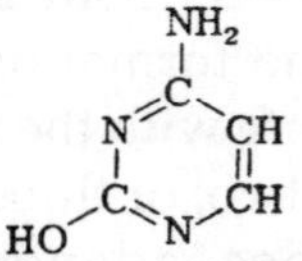

Cytosine (2-(Oxy-4-aminopyrimidine)

Originally DNA was obtained from the thymus because of the high concentration of nuclei in such a gland and was called thymonucleic acid and the RNA was obtained from yeast and was called yeast nucleic acid. It was thus thought that the DNA was characteristic of animals and RNA characteristic of vegetable tissues, however, it has now been found that both DNA and RNA are present in both animal and plant cells but are distributed differently. DNA is concentrated more in the nuclei and this is why DNA was the type of nucleic acid isolated when thymus gland was used as a source.

In the formation of nucleotides, compounds known as nucleosides are first produced. A nucleoside is a molecule which is formed by the combination between a base which can be either a purine or pyrimidine and one or other of the two sugars, i.e., adenine, guanine, cytosine, uracil, or thymine associated with a sugar will produce a nucleoside; if they are associated with the ribose sugar they are one type of compound and with the deoxyribose they form another type of compound. Adenine, for instance, with ribose forms adenosine and with deoxyribose it forms deoxyadenosine; guanine forms guanosine with ribose and deoxyguanosine with deoxyribose; cytosine becomes cytidine with the sugar and deoxycytidine with the deoxy sugar; uracil with the ribose sugar forms uridine but the corresponding compound with the deoxyribose, is not recorded. Thymine forms thymidine with ribose.

Adenosine (9-ß-ribofuranosidoadenine)

The next stage in the formation of a nucleotide is the combination of a phosphate with the nucleoside. For instance, adenosine can have one, two, or three phosphates attached to it and it then becomes adenosine monophosphate, diphosphate, or triphosphate, respectively. Similarly the other compounds mentioned can be mono-, di-, or triphosphorylated. Phosphorylation of the five bases with a single phosphate group forms adenylic, guanylic, cytidylic, uradylic and thymidylic acids and isomers of these are also known. The adenylic acid is commonly spoken of as "muscle" adenylic acid because it was first isolated from muscle. It is one of the structural units of ribonucleic acid and its phosphoric linkage is to the 5-carbon atom. In plants, however, the phosphate linkage in the monophosphate appears to be largely associated with the 3-carbon atom.

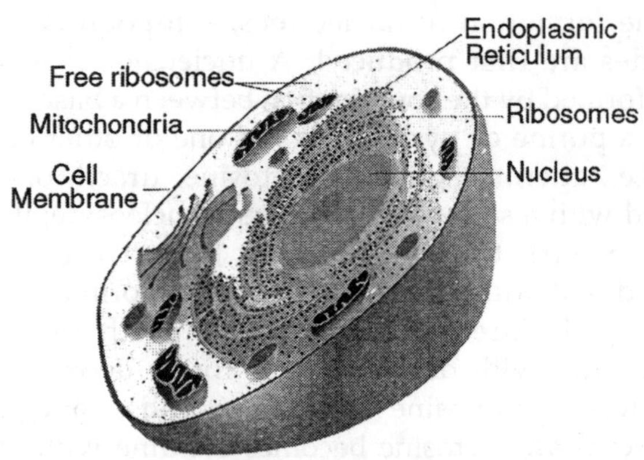

Fig. Function of Cells.

The various nucleotides mentioned in the foregoing become polymerized by the formation of phosphoric acid bonds between the sugars and polynucleotides are thus formed. The common name for these is "nucleic acids." When the nucleotides are made from deoxyribose sugars, the deoxynucleic (also referred to as desoxynucleic) acids (DNA) are formed and if the ribose sugar is involved ribose nucleic acid (RNA) is obtained.

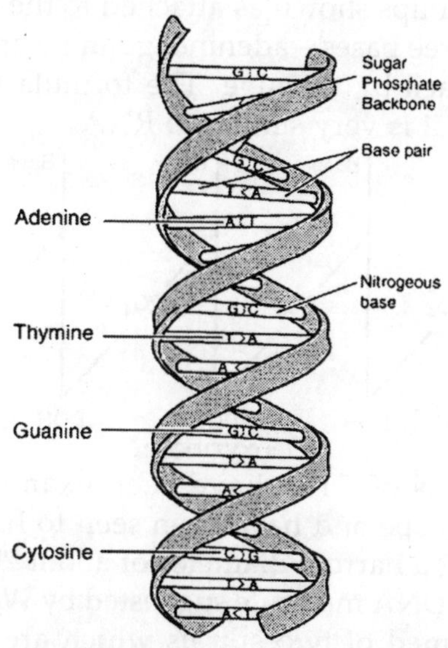

Fig. Model of Molecular Structure of DNA

The structure of these two compounds can be diagrammatically demonstrated,

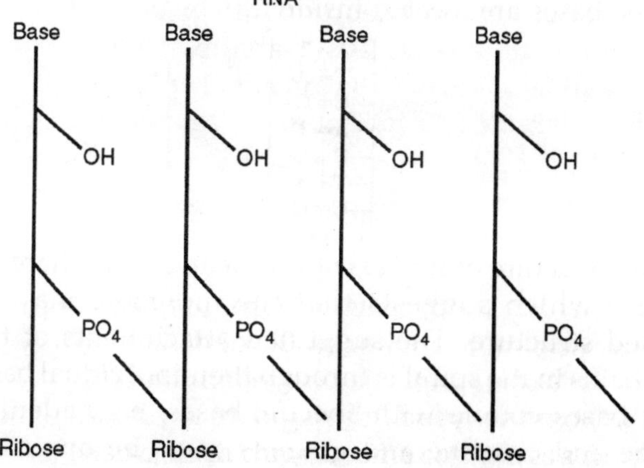

The bases in each nucleotide vary and may be any one of those first mentioned (i.e., adenine, guanine, cytosine, or uracil). The frequency of each in any particular RNA molecule depends upon the origin of the RNA.

DNA has fundamentally the same structure as RNA but lacks the OH groups shown as attached to the sugars in RNA and uses the three bases—adenine, guanine and cytosine but replaces the uracil by thymine. The formula for DNA looks like this, that is it is very similar to RNA.

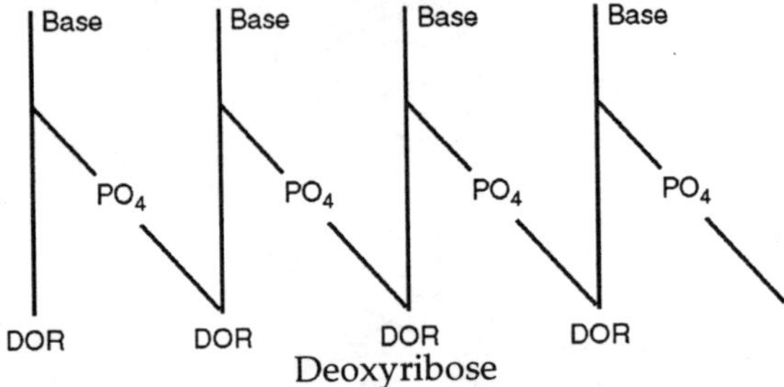

Preparations of DNA have been examined under the electron microscope and have been seen to have the form of long fibers with a narrow diameter of about 20 A across. The structure of the DNA molecule suggested by Watson and Crick is that it is formed of two spirals which are coiled about a common axis and the union between the spirals is by means of hydrogen bonds between the bases of the nucleotides. The phosphates and sugars are actually on the outside of the chain while the bases are located inside it.

A=T
T=A
C≡G
A=T
G≡C
T=A
C≡G
A=T

The structure of RNA is not so clear cut but there is some evidence which suggests that the molecule may have a branched structure. The suggested attachments of the two DNA chains in the spiral is through their individual bases and certain bases couple with specific bases, e.g., adenine and thymine and cytosine and guanine must be opposite each other. The structure therefore is something like this. The order and frequency of bases shown is empirical and would depend on the origin of DNA. We do not yet have a technique for determining the order of bases in the nucleic acids. RNA has not been proved yet to exist as double strands but many attempts are being made to find out for certain whether this is so.

In the living cell the nucleic acids are inevitably combined with proteins to form nucleoproteins and again because of the nature of the sugar one can specify two types of nucleoprotein, deoxyribonucleoprotein and the ribonucleoprotein. The main protein which combines with DNA in these complexes is of a basic nature and is usually either histone or protamine and there are varying amounts of protein and DNA combined under different circumstances. It is of interest that viruses and bacteriophages are largely composed of nucleic acids and in the case of the bacteriophages the DNA can be extruded from the virus particle into the bacterial cell where it can lead to formation of fresh virus but this will be mentioned again later.

NUCLEIC ACIDS OF THE NUCLEUS

DNA is segregated away from the cytoplasm within the nucleus, although, as we have seen, the isolation of the nucleus from the cytoplasm is by no means complete since there are a number of pores which are present in the nuclear membrane and which lead into the cytoplasm as such. But the significant thing about DNA is that practically all of it in the cell is localized in the nucleus and it is the DNA that has the major genetic function in the cell.

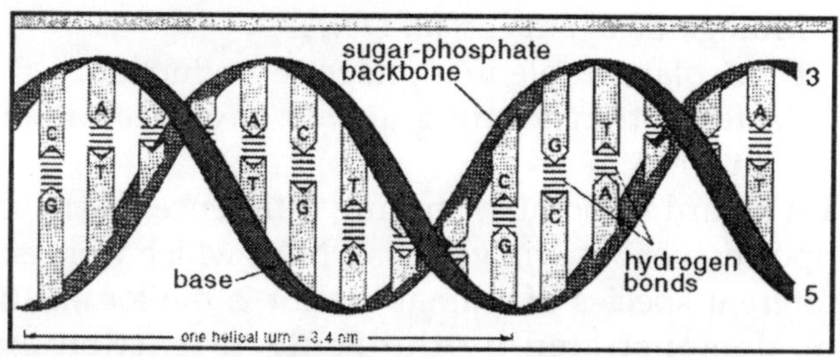

Fig. Section of DNA Helix to show Base-pair Sequences

What evidence do we have that DNA is concerned with heredity? The evidence for this has been excellently presented by Brachet in his book *"Biochemical Cytology"* and will be summarized here. Seven points in evidence for this are:

- There is a specificity of DNA's from different species.
- There is ispecific localization of DNA on chromosomes, e.g., in *Drosophila* there is an identity between the DNA bands and the localization of the genes.
- There is an approximate constancy of amount of DNA per chromosome set.

  RNA and DNA Nucleic acid, which forms part of the composition of viruses, may be present in the form of RNA (ribonucleic acid) or DNA (desoxyribonucleic acid). In general, according their type, the viruses contain only one or other form of nucleic acid; however, examples are known containing both, in the case of the influenza virus. DNA is a tetranucleotide consisting of: H 3 PO 4-R-guanine -- H 3 PO 4-R-cytosine -H 3 PO 4-R-adenine where R indicates desoxyribose, while in the case of ribonucleic acid (RNA) it stands for rebose. At the left is the sytemtic representation of the DNA molecule, form of two pentose-phosphate chains winding helicoidally around a central axis and joined by means of pairs of basis-adenine and thymine, guanine and cytosine. A fraction of the DNA molecule is represented below, with the pair of basis joining the pentose-phosphate rings.
- DNA is relatively metabolically stable.
- Mutagenic agents affect DNA.
- DNA plays a role in a phage reproduction.
- Bacterial transforming agents have been identified with DNA.

With regard to point 1, striking differences were seen in the composition of bases in various DNA's which were isolated from different species of animals. Point 2, the localization of DNA as demonstrated by the Feulgen reaction and its relationship to the genetic areas has been demonstrated. Point

3, some constancy does occur, it is not absolute but there is a marked tendency toward such a constancy. Point 4, DNA is probably the most stable of the phosphorus compounds that are found in the cell and is very much more stable than RNA.

Point 5, all agents which have a mutagenic effect, for example, nitrogen mustard, ultraviolet rays, x-rays, affect DNA. Most of the points quoted so far are circumstantial but point 6, the role of DNA in phage reproduction, seems to provide more direct evidence, as will be explained here. Some investigators, e.g., Hershey and Chase, labeled DNA with P32 (radioactive phosphorus) and its protein with S35 (radioactive sulfur). The phages used (T2 type) have a head and a tail; the membranes of the head are made of protein and contain DNA. When the phage is added to the bacteria which it infects, the tail of the phage becomes attached to the membrane of the bacterium by a submicroscopic sucker.

The tail of the phage has recently been shown to be cross striated and contractile and to contain a solid core. Contraction of the tail shoots the core; like a harpoon, into the bacterium. Attached to the proximal part of the core is one end of a long, single, greatly folded, DNA molecule which occupies the head of the phage. The solid core carries this molecule into the substance of the bacterium just as a harpoon carries with it the rope to which it is attached.

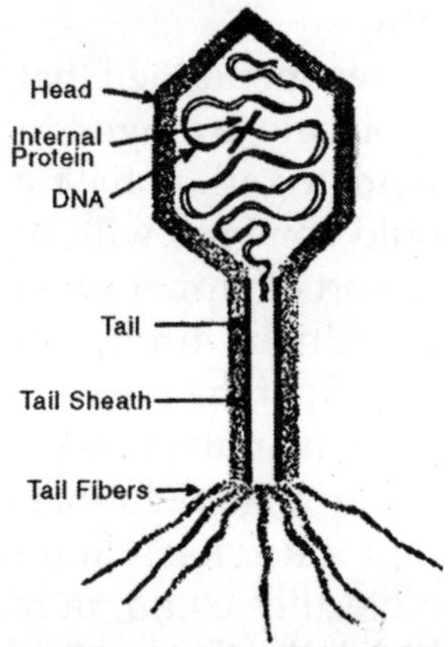

Fig. Infection of Bacteria with T 2 Virus Particles

Practically no protein enters with the DNA and the DNA

then takes over the synthetic process of the bacterial cell and uses its metabolism to produce some virus protein and a good deal of its own DNA instead of bacterial protein. Thus the DNA of the phage is able to produce DNA which is genetically identical to the infecting bacteriophage and genetically quite different from the DNA of the bacterium.

The significance of DNA for bacterial transformations is explained as follows. There are two types of pneumococci one type known as the encapsulated or smooth strain and a second type which is not capsulated and is known as the rough strain. If DNA from the first of these bacteria is brought into contact with the rough strain of Pneumococcus, the latter will be converted into the smooth strain and this property will be transmitted to the descendants of the transformed strains and these descendants will produce DNA with the same properties. Thus the DNA is found to be directly concerned with the synthesis of a particular type of capsule. Bacterial transformations have also been obtained with a number of other bacteria using DNA.

This work has suggested that at least some and possibly all the genes found in bacteria were probably completely or largely composed of DNA. It is of interest that bacterial viruses (bacteriophages) are all DNA viruses, that plant viruses are all RNA viruses and that some human viruses contain RNA and some contain DNA.

It has previously been mentioned that the model for DNA was a two stranded molecule arranged in a helical form with the pairs of bases held together by hydrogen bonds, for example, adenine would combine with thymine and guanine with cytosine. This is a sort of zipper arrangement and if these pairs of molecules could be unzipped by breaking the hydrogen bonds, then each of the members of the chain could serve as a template for the manufacture of a second chain. This is the type of molecular arrangement that would be necessary to enable a gene to reproduce itself and there is very strong evidence that DNA molecules could, in fact, be equated with genes. Although DNA fulfills these requirements, we cannot be sure, of course, that when it is incorporated into the cell it simply exists as DNA. Most likely it combines with a protein which probably exerts an influence on the DNA in its function as a gene.

Although the synthesis of protein in a cell is attributed largely to RNA, there is some evidence that DNA is concerned with this process and that it is particularly concerned with the sequence of the arrangement of amino acids in the protein being synthesized; but it is thought that even this activity on the part of DNA is mediated by RNA. In experiments in which bacteria were disrupted, it was found that incorporation of amino acids into protein required the presence of DNA before it could take place but that RNA also appeared to be necessary.

One very difficult and complex problem to explain is the mechanism by which the gene, if it is composed largely or exclusively of DNA, is able to pass on the genetic information which it contains so that the various cells conform to the genetic requirements. In this connection the reader should study Danielli's interesting (discussion on the theory of nucleocytoplasmic relationship. The important thing about the DNA molecule is not so much its length as the sequence of nucleotides in it. This information must be transferred and it is the proteins to which this transfer is made. Probably what happens is that the DNA molecule and its sequence of nucleotides decide the sequence of amino acids in the proteins which are being built up. It is thought that this is not done directly but that the information is first transferred from the DNA to the RNA and it is the RNA which serves as the template.

The problem of the relation of RNA to protein synthesis will be discussed shortly. It appears possible that each region or nucleotide sited at a particular point on the RNA template binds a specific amino acid and the sequence of these amino acids is determined by the sequence of the sites on the RNA. The intensity of these reactions is noteworthy, genetic effects are well known to be most resistant to unstable conditions, particularly environmental disorders. Genetic information tends to exert its effect despite the subjection of the organism concerned to variations in temperature, salinity of the fluids with which it comes in contact, different types of food, the amount of light it receives, exposure to toxic substances and so on. However, radiation is one environmental factor which can exert it profound effect on the genetic material.

An interesting thing concerning the passing on of genetic information to the organism by the DNA is that there is no feedback of information to inform the DNA, so far as we know, of what has been done or what needs to be done but, nevertheless, practically all significant cell activity is affected by the gene. It is certainly a very complex physicochemical problem and there is still a very long way to go before it can be adequately understood. Most interesting from the point of view of the subject of this book is the fact that the DNA, tucked away in the nucleus as it is, can exert its effect throughout the cell, not only determining the type of protein which is laid down in the cell during the process of development but controlling a large proportion of the activities of the cell in the adult animal.

RIBOSE NUCLEIC ACID (RNA)

Just as DNA is thought to carry information concerning the rice, it has been suggested recently that RNA is concerned with carrying information concerning the individual, in other words the RNA molecules are concerned with the process of storing information. It has in fact been suggested that memory is stored in the RNA of the neurons of the brain. This, however, is largely theoretical at the moment but is an intriguing prospect for further investigation.

There seems now to be extremely good evidence that RNA is concerned with the synthesis of protein. Before we go on to consider this perhaps we should say a word about the nature of protein.

If proteins are hydrolyzed they break down to amino acids and under natural circumstances a protein is made up of amino acids joined together in the form of elongated chains known as polypeptide chains. In a protein of molecular weight of about 25,000 there will be, according to F.H.C. Crick, something like 230 residues joined end-to-end to form the single polypeptide chain of which the molecule is made. The amino acids are invariably joined together in this chain by the same method. The bonds are covalent bonds and the amino acids are joined together by the formation of a peptide link which results in the elimination of a molecule of water; nearly all covalent links in a protein are formed by this method, however, occasionally there are links between two sulfur molecules.

The protein so formed therefore is an elongated linear molecule and Crick points out that there is no evidence that it is branched. Crick also makes the point that there are only about 20 different kinds of amino acids in proteins but these same 20 occur, generally speaking, in practically all proteins--it does not matter whether the proteins come from animals or plants or microorganisms. He points out, however, that not all proteins contain all these amino acids but the majority of proteins do contain a proportion of them. Another point is that all the amino acids present in proteins have the L-configuration. There are some specific amino acids which occur only in specific proteins and probably the best example of these is hydroxyproline, the amino acid which is characteristically found in collagen. The twenty amino acids found universally in proteins are as follows: glycine, alanine, valine, leucine, isoleucine, proline, phenylalanine, tyrosine, serine, threonine, asparagine, glutamine, aspartic acid, glutamic acid, arginine, lysine, histidine, tryptophane, cysteine, methionine. Other amino acids which may be present are: hydroxyproline, hydroxylysine, phosphoserine, diaminotumelic acid, thyroxine and cystine. Crick points out in his article that not only is the composition of a particular protein fixed but it appears that the order of the amino acids in the polypeptide chain is determined very exactly and that, for example, every molecule of hemoglobin in human blood has precisely the same sequence of amino acids as every other molecule of hemoglobin.

A further point is that the polypeptide chains of the proteins are not extended but are folded upon each other and this folding is said to be maintained by physical bonds which are weak in nature and possibly also by disulfide linkages and other linkages may exist as well. The degree of folding is thought to be similar for each particular protein and presumably the denaturation of proteins results in the destruction of this folding. If any one of the twenty essential amino acids is supplied to a cell, it can be incorporated into proteins but if, for example, any one of the twenty is not available to an organism, protein synthesis cannot occur.

Not only is the synthesis of that part of the protein molecule which does contain the specific amino acid prevented but also that part of the molecule which does not contain the amino acid is also not synthesized. Crick points out that, with this type of mechanism and with the order of the amino acids being very specific, one would expect that some variation or mistakes in order would occur occasionally, but as he points out the information we have at the moment is that such mistakes are very infrequent.

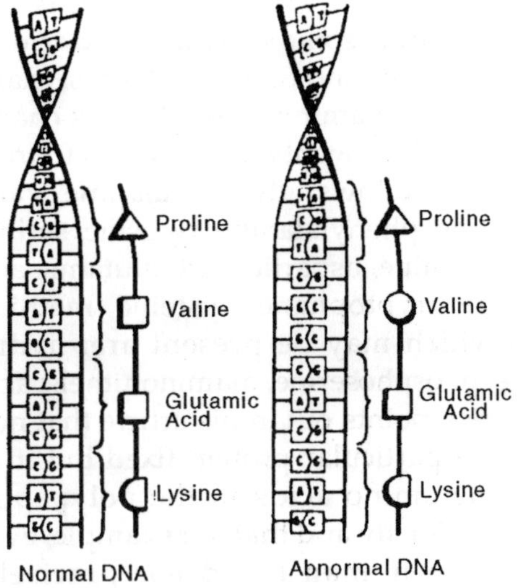

Fig. Synthesis of Normal and Sickle cell Hemoglobin

Although all the hemoglobin molecules of the human being are the same, if this hemoglobin is compared with that of the horse or some other animal, it will be found that there is a general similarity of pattern between the two molecules and that the amino acid composition will be pretty much the same for both of them. These two hemoglobin molecules may differ a little in their electrophoretic properties--their crystalline form may differ and the ends on their polypeptide chains may be different, but it is very likely that the sequence of amino acids in the polypeptide chains will be fundamentally the same except for one or two slight alterations in the sequence.

It is of interest that this produces what might be described as a family likeness between proteins and Crick suggests that it may be that "these sequences are the most delicate expression possible of the phenotype of an organism and that vast amounts of evolutionary information may be hidden away within them." It seems possible and this will be further discussed, that the sequence of amino acids in the proteins is under control of Mendelian genes and the classic case of this has, in fact, been described by Pauling and his co-workers in 1949.

It was found that the nature of the protein in the hemoglobin in human sickle-celled anemia is different from that of normal humans. This disease has been described by Pauling as a "molecular disease." Ingram, has shown that this difference results from the fact that valine replaces glutamic acid in the chain and this, according to Ingram, is the only change present in the molecule. Crick says of this "it may surprise the reader that the alteration of one amino acid out of a total of about 300 can produce a molecule which (when homozygous) is usually lethal before adult life."

When the process of synthesis of proteins is studied, one particularly important problem has to be considered and that is how to explain the mechanism which controls the order of the amino acids in a protein. The condensation of amino acids into polypeptide chains is fairly simple to explain chemically but their order is not. This order must be very critically and stringently controlled since, as we have seen in the foregoing, the slightest variation in sequence of amino acids in the hemoglobin molecule can produce a lethal disease. Thus the critical point in protein synthesis is the joining up of the amino acids in a predetermined order. Now, it has been mentioned earlier in this discussion that nucleic acids are concerned with the synthesis of protein and it is almost certain that DNA has an important influence in deciding the sequence of amino acids. That DNA can affect protein synthesis is demonstrated by the fact that when DNA is squirted by the T2 bacteriophages previously described into bacterial cells without any of the protein going in with it, it appears to be able to control the synthesis of protein inside the bacterium.

PERMEABILITY OF THE NUCLEAR MEMBRANE

If the nucleus is concerned with protein metabolism, we should say a word about the penetration of the nuclear membrane, because in animal and plant cells protein or some other substance must get out into the cytoplasm for the nucleus to exercise control over protein synthesis. We have mentioned before the effect which the existence of pores might have on the permeability of the nuclear membrane. Some direct permeability studies have been carried out and, while there is no intention here of trying to review this work, it may be pointed out that polypeptides and proteins seem to penetrate into the membrane readily. Conversely, dipeptidase and other enzymes have been shown to diffuse out readily from isolated nuclei.

However, the permeability of the membrane in such cases may result from damage to the nuclei during the process of isolation. Brachet has described experiments in which both ribonuclease and deoxyribonuclease appear to penetrate the membrane of isolated nuclei and to cause an alteration in its shape. The same criticism which has just made could apply to his work, but Brachet anticipated this by demonstrating the penetration of ribonuclease (molecular weight 13,000) into the nuclei in intact amoebae and onion tip cells. Protamines and histories also appear to penetrate readily into the nuclei of living cells and this process according to Fischer and Wagner requires energy since it is inhibited by dinitrophenol and cyanide. These few facts support the conception that the nuclear membrane serves only partly to isolate nuclear activity from that of the cytoplasm. Further information about the permeability of the nuclear membrane can be obtained from Brachet book "Biochemical Cytology."

Up to the present we have been considering whether protein is being synthesized only in the nucleus, we know that both DNA and RNA are present in the nucleus and that DNA exerts an effect on protein synthesis. Perhaps we should divert for a minute here to consider what is probably one of the most important pieces of evidence of the relationship of RNA to protein production. This is the work which has been carried out recently on tobacco mosaic virus by two groups, by the Gierer and Schramm group and by the group of Fraenkel Conrad. There are a number of strains of tobacco mosaic virus and some of these strains have a protein which contains histidine. Gierer and Schramm showed that if you take the RNA part of the virus and free it completely from protein, it can still be infective although it is not as infective as when combined with protein.

Fraenkel Conrad was able to recombine the protein and RNA of the virus and obtain again an active virus in which the infectivity was high but not quite normal. It is of interest that by swapping the RNA's and proteins from different strains it is possible to obtain different combinations. If, for instance, the virus, which is made by taking the RNA of one strain and the protein from another, is used to infect a tobacco plant, the virus which is reproduced in the plant is similar not to the virus from which the protein was taken but to the one from which the RNA was taken. So that, for example, if in the particular strain of virus from which the RNA was taken there was no histidine in the protein, then the virus removed from the tobacco plant after infection would also have no histidine; in other words, the RNA of the virus determines the composition of the protein of the virus. Furthermore the protein which was in the original virus was reproduced in the plant and this protein was that of the virus from which the RNA originally came.

Let us return now, however, to the question of where in the cell is protein produced? There is RNA present in the nucleolus and RNA present in the form of small granules about 150 A in diameter associated with the endoplasmic reticulum. This endoplasmic reticulum, as has been described previously, breaks up on homogenization of the cell and is centrifuged down as microsomal particles. Crick makes the following comment: "Biologists should contrast the older concept of microsomes with the more recent and significant one of microsomal particles. Microsomes came in all sizes and were irregular in composition, microsomal particles occur in a few sizes only, have a fixed composition and a much higher proportion of RNA. It is hard to identify microsomes in all cells whereas RNA-rich particles appear to occur in almost every kind of cell. In short, microsomes were rather a mess whereas microsomal particles appeal immediately to one's imagination and it will be surprising if they do not prove to be of fundamental importance."

SYNTHESIS OF PROTEIN BY NUCLEOLI

We shall first discuss briefly the case for the synthesis of protein by the nucleoli. It has already been mentioned that the nucleolus contains a good deal of RNA but it also contains some DNA and Caspersson claims it is rich in diamino acids. The presence of protein masked phospholipids has also been claimed and alkaline glycerophosphatase and acid phosphatases have been demonstrated in this organelle by a number of authors and also a great variety of other phosphatases by the present author. Other enzymes are present in the nucleolus, e.g., dipeptidase, cytochrome c reductase, nucleoside phosphorylase and the DPN synthesizing enzyme, the two latter are present in a higher concentration in the nucleoli than in the cytoplasm. Many SS-- and SH-- groups are also present and, in general, the nucleolus has from about 40 to 85% of dry matter, which is, relatively), speaking, a very considerable amount.

Although the nucleolus is an area in the nucleus where there is a large concentration of special substances, the electron microscope has to date given no evidence that it is surrounded by a limiting membrane. Originally Caspersson suggested that the nucleolus was the centre of protein synthesis and part of this suggestion is due to the fact that the nucleolus is large and well developed in protein synthesizing cells and that in such cells it is particularly rich in RNA.

The conception that the nucleolus plays a part in protein metabolism has been made likely by the work of Ficq. She noted by tract autoradiography that glycine and phenylalanine containing radioactive carbon are incorporated more rapidly into the nucleoli than into the cytoplasm. In the case of oocytes which are growing rapidly and thus synthesizing much protein the rate is extremely fast. Similarly nucleic acid precursors are concentrated from 100 to 1000 times faster in nucleoli of such eggs than in the cytoplasm.

Further mention should be made of the localization of alkaline phosphatase in nucleoli. This was first recorded by the present author in 1943 as a result of histochemical studies and was subsequently confirmed by a number of workerssince then the present author has described the remarkable dephosphorylating activity of the nucleoli for a wide variety of phosphate substrates. This is shown particularly by the Purkinje cells of the cerebellum and by other cells which produce appreciable amounts of protein. The nucleoli of these cells readily dephosphorylate glycerophosphate, numerous sugar phosphates, riboflavin-5-phosphate, pyridoxal phosphate, ethanolamine phosphate, various steroid phosphates and a wide range of phosphate esters which are concerned with intermediary metabolism and particularly the high-energy phosphates.

In cells which are actively producing protein, phosphatase is not only present in the large nucleoli but in the cytoplasm of the cell as well. Bradfield, for example, showed that an active phosphatase is present in the cells of the silk spinning glands of insects which synthesize and secrete protein at an extraordinarily fast rate. Phosphatase is also present in the cells and fibers during proliferation of connective tissue after wounding. There is an increase in the enzyme in the fibre-producing cells at the time of fibre production. In periosteal fibrosarcomas where no bone is produced and also in polyostotic fibrous dysplasia, alkaline phosphatase is seen in both nucleoli and cytoplasm of the fibre-producing cells.

Siffert has pointed out that the frequent association of alkaline phosphatase with the matrix of both fibers and cartilage would indicate an association with matrix production. It is of interest that phosphatases are present in the uterus of the hen and the mantle edge (which secretes the shell) of mollusks. Both the uterus and the mantle edge produce considerable amounts of protein. The egg shell of the hen has no calcium phosphate in it and the shell of the mollusk has only very small amounts of phosphate so that it seems in these cases the enzyme is more concerned with matrix production, in other words with protein synthesis than with mineralization; the enzyme is also present particularly in regions where histogenesis is occurring. Jeener has shown that alkaline phosphatases are associated with cell proliferation in organs stimulated by sex hormones. Some phosphatases are also present in other parts of the nucleus but their significance will not be discussed here.

These facts together with the association of the enzymes with the nucleolus strongly support the concept of the importance of the latter in protein synthesis. It is of interest that here is a localized region of the nucleus without an obvious membrane in which, presumably, the protein metabolism of the nucleus is centered. Although the nucleolus is itself separated from the cytoplasm by the surrounding nuclear elements, as mentioned earlier, it is mobile and able to make direct physical contact with the nuclear membrane on occasions. It is possible, that on these occasions the nucleolus may discharge its protein through the nuclear membrane. Whatever the role of the nucleolus in protein synthesis, we must accept the conception that a substantial amount of protein synthesis also takes place in the cytoplasm.

Cytoplasmic localization of RNA provides some evidence for this. The recent work of Barrows and Chow on the intracellular distribution of vitamin B 12 adds support to it. Vitamin B 12 is associated with protein synthesis possibly through the formation of RNA, but RNA and DNA are decreased in vitamin B 12 deficiency and the incorporation of P32 into nucleic acids is effected in this condition. The distribution of vitamin B 12 in liomogenates of the liver is 40% in the microsomes, 13% in the mitochondria and 11% in the nuclei, with 22%, in the supernatant. Although only 11% of vitamin B 12 is present in the nuclei, if this were all localized in the nucleolus it would represent an appreciable concentration in this region.

It is of interest in view of what has just been said about the nucleolus that protein synthesis can, in fact, take place in the cytoplasm even in the absence of the nucleus. It appears that the RNA particles which are associated with the endoplasmic reticulum are the centre of such activity and it is noteworthy that in the synthesis of protein the amino acids have been said to "flow" through these particles.

They showed that if rats were given fairly large quantities of radioactive amino acids, later, after the animal was killed and liver homogenates prepared, the microsomal particles (the endoplasmic reticulum) contained a constant amount of this particular amino acid indicating that they were fixed by this part of the cell.

However, in the second experiment a very small amount of radioactive amino acid was administered and, in this instance within a short time, the radioactivity present in the microsomes rose steeply and then fell away quite quickly. This is obviously what one would expect if the protein in these particles was turning over very rapidly.

Whether this in fact means that the amino acids are adsorbed onto the microsomes, converted into proteins and then passed out of the microsomes again as protein one cannot say but this is a distinct possibility.

There seems to be some evidence now that the amino acids have to be activated before they can be condensed into proteins and this activation requires the presence of ATP. An amino acid-activating enzyme has now been discovered and found to be widely distributed and is believed to be present in all cells which undertake the synthesis of protein. Following the activation of amino acids, the next step in protein production appears to be the transfer of the amino acids at least in the cytoplasm to the RNA particles of the endoplasmic reticulum by what appears to be a covalent bond. If the amino acids are labeled with radioactive atoms, then the RNA becomes labeled and can be extracted from the solution, purified and then added to the microsomal fraction of a homogenate. The labeled amino acid is then transferred from the RNA to the protein of the microsome.

This is an extremely interesting step in the problem of protein synthesis. It is also of interest that RNA can be synthesized without protein necessarily being synthesized at the same time. Normally one finds that in a cell which is undergoing active RNA synthesis, active protein synthesis is taking place at the same time. However, the synthesis of protein in such a system can be brought to a full stop by the use of chloramphenicols, this is particularly well shown in systems obtained from bacteria.

However, if the protein synthesis is stopped in this way by chloramphencol, synthesis of RNA continues unaffected. In the synthesis of protein by certain bacteria, e.g, Escherichia coli (Gross and Gross, 1956) there are some mutants of this organism which require a specific amino acid. If this amino acid is not supplied then the synthesis of both protein and RNA comes to a stop. If the chloramphenicol is given nothing further happens but, if in the next stage a small amount of the required amino acid is added, then RNA synthesis starts up very rapidly without affecting the protein synthesis.

Then, if the chloramphenicol is removed from the system, protein synthesis will start again and proceed very rapidly. Crick in his article in 1958, points out that the concepts of protein synthesis include two important points.

- The specificity of a nucleic acid is expressed solely by the sequence of its bases and this sequence is a simple code for the amino acid sequence of a particular protein in which the nucleic acid is concerned in synthesis.
- Information once passed into a protein cannot get out again (the central dogma).

In other words, the transfer of information from nucleic acid to nucleic acid or from nucleic acid to protein may be possible. (By information we mean here the sequence of nucleotides in the case of nucleic acid and of amino acids in the case of proteins.) However, the transfer of information in reverse, that is, the sequence of units from protein to nucleic acid or from protein to protein is impossible. Not everybody agrees with this suggestion and further information is required about it.

One should also mention that an appreciable amount of RNA is present in mitochondria and it is possible that mitochondria are capable of synthesizing protein. Studies in the author's laboratory by Sheridan have already been mentioned in which it was found that, in the liver cells of guinea pigs which were subjected to scurvy, the mitochondria were surrounded by layers and layers of ergastoplasmic reticulum or membranes, almost as if these were being synthesized on the surface of the mitochondria.

In fact they are so closely applied to the surface of the mitochondria in some cases that it is very difficult to distinguish them from the mitochondrial membrane. Whether this means, in fact, that there is synthesis of this membrane going on or whether it means that the membrane has become closely apposed to the mitochondria, because the energy derived from mitochondria is required in the process of protein synthesis as carried out by the endoplasmic reticulum, one cannot tell. There is also a possibility that, since in scurvy protein synthesis is depressed, the mitochondria may be synthesizing additional endoplasmic reticular membranes in an attempt to compensate for the depressing effects of the vitamin C deficiency. Mitochondria are enormously increased in numbers in scurvy and also in plain starvation.

It is of interest that the RNA particles of the endoplasmic reticulum like those of the nucleoli measure approximately 150 A across and hence uniformity of size as well as uniformity of structure may play a part in protein synthesis. We have already mentioned that the synthesis of RNA which itself determines synthesis of protein is probably under the control of the DNA, i.e., by the nucleus. This we know, for example, because a Mendelian gene controls the sequence of amino acids in human hemoglobin as shown by the work on sickle cell anemia and furthermore, spermatozoa transfer only DNA and no RNA.

What is the significance of the protein part of the RNA particle we do not know, presumably it is structural though it may be enzymic in nature. Thus, at the moment we can only agree with Crick's comment "the RNA forms the template and the protein supports and protects the RNA." Crick has suggested that the ribonucleoprotein particle is an open structure like a sponge and possibly molecules of appropriate size can diffuse in and out of it, however, the whole relationship of RNA to the protein in these particles requires further study.

The really critical problem in protein synthesis is actually, since there are only four bases in the nucleotides, what arrangement of four bases in the RNA molecule could determine the sequence of twenty amino acids in the synthesis of a protein. This problem is described as the "coding problem" and is one in which the next steps in our understanding of protein synthesis might be taken. There is very little experimental work in this direction as yet and most of the discussion has been hypothetical.

NUCLEOCYTOPLASMIC RELATIONSHIPS

This leads us on to the problem of the relationship between the nucleus and the cytoplasm and their mutual interreactions. In the first place we should consider such reactions in unicellular organisms. Full details of this subject should be obtained by reading Brachet "Biochemical Cytology." Only a summary of these very interesting facts can be given here.

We know that the nucleus is largely composed of DNA, RNA, histone and a number of proteins. The RNA is concentrated exclusively in the nucleoli and is very labile metabolically and this is one of the reasons why it has been suggested that nucleolar RNA may be the source of the RNA of the cytoplasm. However, there are reasons to believe that, although it may contribute some part of the cytoplasmic RNA, which would be in agreement with Caspersson's observations, it does not contribute all of it and that the cytoplasm itself is the site of an appreciable synthesis.

Both histochemically and biochemically it has been demonstrated that nuclei contain extremely little of the oxidative enzymes (cytochrome oxidase and succinic dehydrogenase) which are so characteristic of mitochondria. So that presumably the chemical events which take place in the nucleus are those which are predominantly anaerobic in nature. Nuclei contain glycolytic enzymes and also an enzyme which synthesizes DPN, using nicotinamide, nucleotide and ATP. There are also present a number of enzymes which are concerned in purine and nucleoside metabolism. These occur in greater quantity in the nucleus than in the cytoplasm, at least as far as liver cells are concerned. It has been suggested by various authors, including Brachet, that these findings are in agreement with the fact that the nucleus could be the site of nucleotide, coenzyme and nucleic acid synthesis.

Many of the experiments which have endeavored to show some relationship between the various cellular components have been carried out on homogenates or by mixing of, say, nuclear fractions and mitochondrial fractions and so on, but these are likely to give a very misleading idea of what in fact does happen in the living cell. How important this is, is demonstrated by very interesting experiments carried out by de Fonbrune in 1939. He transplanted nuclei from one amoeba to another and showed that the nuclei control the characteristic streaming of the cytoplasm in each case so that the type of streaming of the cytoplasm of one amoeba is transferred to another species of Amoeba when the nucleus of the first cell is pushed into the second.

The point of these experiments however, is that in the process of transfer the cell walls of the two amoebae must be in extremely close contact with each other and the nuclei are then simply pushed through the two cytoplasmic walls by means of a blunt glass probe. If, however, the nucleus is pushed out into the medium and then grafted back into the cell by pushing into the other cell, it loses its ability to divide and, in fact, many of its activities seem to stop. So one can only guess at what happens to the nuclei and mitochondria and possibly other parts of the cells when the latter is homogenized, washed in sucrose, spun down in centrifuges, drawn up in pipettes, squirted out of the pipettes, mixed up with various reagents and so on.

It is pretty certain that the cell organelles being used in this way are in a different state from those that exist in the living cell. It is interesting that Cutter and his colleagues have demonstrated that coconut milk contains a number of nuclei which swim freely in it and which presumably come from the endosperm cells. These nuclei, which are in a physiological liquid that the coconut supplies, are still not capable of mitosis and are capable only of degenerative division when they are transplanted back into endosperm cells.

Studies have been carried out on the Amoeba and on the unicellular alga, Acetabularia, which involve sectioning the cytoplasm so that one-half contains a nucleus and the other does not. The groups of workers concerned with these studies are Hammerling and his colleagues, Mazia and Danielli and co-workers.

It cannot engulf living organisms but, if both the nucleate half of the amoeba and the non-nucleate half are kept starving, the nucleate half dies within about three weeks and the half without a nucleus in two weeks so that is about 50% longer survival resulting, presumably, from the presence of the nucleus. If the nucleus is grafted into a fragment of cytoplasm of the amoeba that has no nucleus and if the animal has only been in this condition for two or three days, then there is a very dramatic reintroduction of pseudopod formation and characteristic amoeboid motility. However, if the half has been without a nucleus for about a week or longer then the introduction of a nucleus does not have the same rejuvenating effect and it is apparent that irreversible changes have taken place in the cytoplasm.

Danielli has expressed the opinion that particularly in these amoebae and possibly in cells of higher animals the nucleus is responsible for the type of macromolecules which are produced whereas the cytoplasm plays the part of organization of these macromolecules into what he describes as functional units. On the other hand, there are probably also mutual interreactions between the cytoplasm and the nucleus; it is not just all one-way control.

Hammerling has carried out a number of extremely interesting experiments on Acetabularia and we will take this opportunity of referring here to some of them. Acetabularia, a unicellular alga, normally has a stalk which contains some chloroplasts and a series of rhizoids and in one of the larger rhizoids the nucleus is situated.

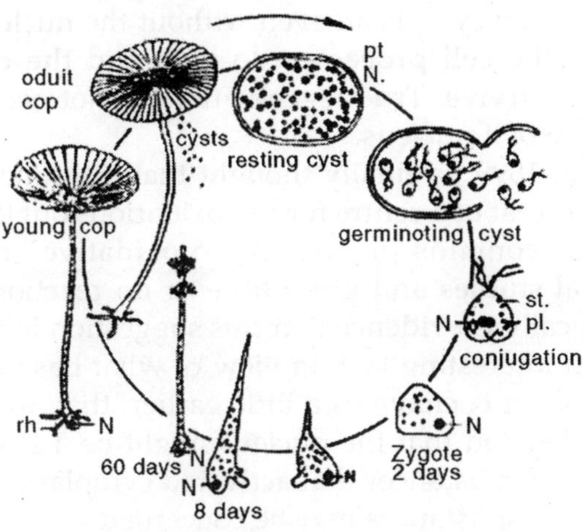

Fig. Life Cycle of Acetabularia Mediterranea

At an appropriate period, the stalk forms a cap, which looks rather like that of a mushroom and produces cysts which germinate and undergo a sexual union to form a zygote that reproduces the stalk and the rhizoids again. If a second rhizoid containing a nucleus is grafted onto the stalk of an acetabularian, it will form two of these caps instead of one. Here is an example of the effect of the nucleus on the cytoplasm. In the reverse experiment, Hammerling showed that, if a cap is removed from the acetabularia just before the nucleus goes into division, then nuclear division will stop and will not occur until a new cap has been formed.

Presumably something necessary for the nucleus to divide is secreted by the cap and, if the cap is continually removed, the process of mitosis can be delayed indefinitely. On the other hand, if the nucleate part of a rhizoid is grafted onto a plant which already has a nucleus, then a cap nuclear division can be produced within two weeks instead of what would be the normal time of about two months. In Spirogyra it was shown 50 years ago that, if the nucleus is removed, the non-nuclear parts of the cell survived for quite a long time, carried out photosynthesis, formation of plasticis and fats and production of tannic acid; the protoplasm continued to stream and there was even an increase in length. Thus a good deal of activity can go on in the cytoplasm even without the nucleus and yet in the end the cell processes do stop and the cell cannot continue to survive. True regeneration cannot occur without the presence of a nucleus.

Loeb in 1899 originally thought that the function of the nucleus was that of a centre for cell oxidations but the fact that the nucleus contains practically no oxidative enzymes by biochemical studies and gives little or no reaction for these histochemically is evidence that this suggestion is not correct.

It is an interesting fact, in view of what has been said of the synthesis of coenzymes a little earlier, that, in 1925, E. B. Wilson had stated that the nucleus might be a storehouse of enzymes or of substances that activated cytoplasmic enzymes and that these substances may be concerned with synthesis as well as with destructive processes. It was suggested also, in 1892, by Verworn that the nucleus was the main synthetic centre of the cell. This suggestion was also supported by Caspersson in the light of his various studies and he thought that the nucleus was the principal centre for protein formation in the cell. Mazia in 1952 suggested that the function of the nucleus is really that of a replacement of products of cell activities and in support of this he points out that removal of the nucleus is not followed by effects which take place immediately but instead they take place over a period of time.

For example, the nucleus produces enzymes, if the nucleus is removed then the cytoplasm will continue to function with the enzymes it has until they drop below a functional level--this may take place at different rates for different enzymes. Mazia has also suggested that as an alternative the nucleus might produce what could be described as a cytoplasmic unit which would be comparable to a plasmagene. It would play a part in cytoplasmic synthesis and would have to be replaced all the time by the nucleus if the cytoplasm was to be maintained normally. As soon as these cytoplasmic units were exhausted then the cytoplasm would not carry on its normal activities. Thus, although the nucleus may not be the centre of oxidative processes in the cell, it may influence these processes and we should consider the evidence for and against this.

Shapiro in 1935 by the process of micrurgy cut sea urchin eggs into two pieces, one half contained the nucleus and one did not and he found that the oxygen consumption was much higher in the fragment which contained no nucleus. This is as one would expect from our knowledge that the cytoplasm contains mitochondria and that the mitochondria contain most of the respiratory enzymes of the cell. Brachet has also studied the oxidative processes in Amoeba proteus and he found that the removal of a nucleus had very little effect on the rate of respiration and, in fact, there was no change for at least seven days, but after that time the cytoplasmic fragments began to undergo cytolysis and the respiratory. rate then fell off.

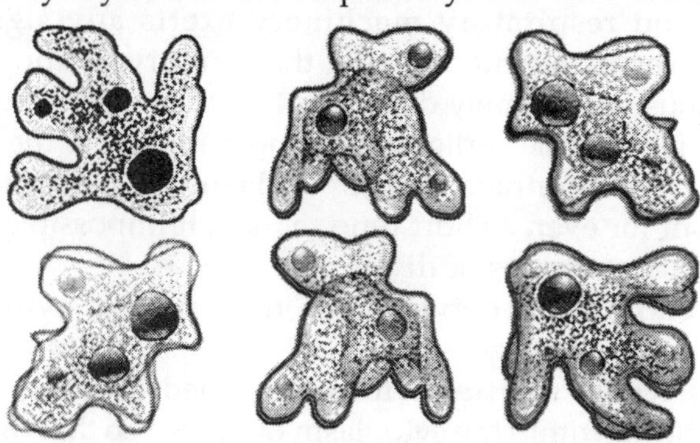

Fig. Amoebae with Severed Portions

The drop in the latter is understandable since the maintenance of the proper structural organization of the mitochondria and probably the related cytoplasm is necessary for the respiratory rate to be maintained. In the case of Acetabularia a similar result has been obtained, however, some studies which were carried out by Whiteley on the protozoan, Stentor, are of interest in this connection.

He too severed his protozoan into two portions, one of which contained the nucleus and one did not and showed that in the fragment with. out a nucleus there was a drop in the consumption of oxygen and no regeneration, however, in the fragments which had a nucleus, regeneration occurred after 24 hrs. and there was a corresponding increase in 02 consumption during this period. Eventually the respiratory rate came back to normal.

These results suggest that the macronucleus has some control, perhaps not over the actual respiration but over the synthesis of fresh respiratory equipment, in other words, of fresh mitochondria. The results with the amoebae and with Acetabularia, as Brachet says, completely invalidate Loeb's older conception that the nucleus was the centre of oxidation processes in the cell. In fact, it is quite striking that the oxygen uptake is unaffected for quite a long time after the nucleus is removed from cells--in the case of the amoeba this is as long as ten days, in the case of Acetabularia as long as three months--and one cannot conclude from this that the nucleus once it has formed respiratory machinery exerts any significant control over its functioning. On the contrary, it appears that the nucleus is extremely dependent on the cytoplasm and we have noticed a little earlier how important it is for the nucleus to be kept in contact with cytoplasm. Its removal from cytoplasm for even a short time renders it impossible for it to carry on its processes of division.

A relationship between the nucleus and cytoplasmic components is suggested by the results of tissue culture workers who have frequently described the migration of mitochondria within the cytoplasm of the cell so that they come into contact with the nuclear membrane, remain there for some time and finally disengage themselves. The nucleolus has often been observed to move through the nucleus and touch the nuclear membrane on the inside. It has been mentioned that usually it is met there by a mitochondrion. The cyclic movements of nucleolus and mitochondria in spinal ganglion cells noted by Tewari and the present author have also been already mentioned. In these cases it is possible that the nucleolus is discharging RNA or DPN through the nuclear membrane or, in the case where the mitochondrion is associated with the cell membrane, ATP is perhaps passed inward to the nucleolus.

Although the nucleus does not appear to have any special relationship to respiratory activity in the cytoplasm, it does have other effects of a quite fundamental nature. This has been very well demonstrated by experiments which have been carried out on amoebae by Brachet and his colleagues and which have given extremely interesting results. They showed that if amoebae are cut into nucleate and non nucleate parts, although the oxygen consumption is not significantly affected by this process, there is a disappearance of the ability to utilize glycogen in the nonnucleate part. In other words, the glycogen stores are scarcely affected, but, on the other hand, the protein tends to be used up so that the nucleus appears to control the process of glycogenolysis and also maintains the integrity of the protein of the cell.

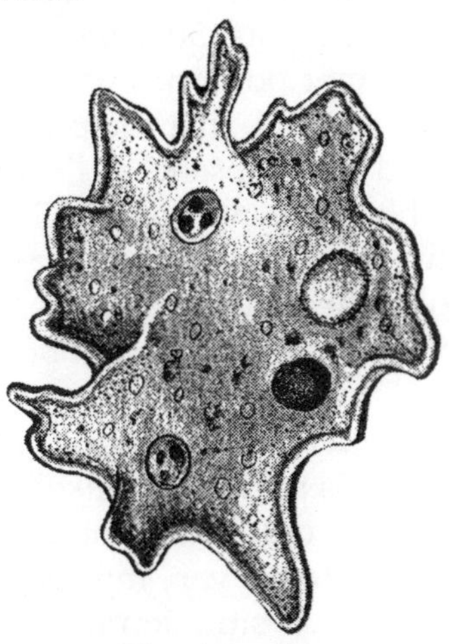

Fig. Amoebae

It is of interest that it takes a period of about 3 days before the carbohydrate breakdown stops and protein catabolism becomes increased. It is obvious, therefore, that the nucleus exercises very profound control over fundamental processes in the metabolism of the cytoplasm. Although there is a lag period before changes in the glycolytic system and protein metabolism, occur in the nonnucleate part of the amoeba, changes in phosphate metabolism take place almost instantaneously. It was originally demonstrated by Mazia and Hirschfield in 1950 that the uptake of radioactive phosphate by the nonnucleated part of the amoeba was substantially less than that of the part containing the nucleus.

Brachet and his colleagues repeated this experiment, first they placed whole amoebae in radioactive phosphate for some time and then separated the amoebae into nucleated and nonnucleated parts and demonstrated that there was no particular uptake of the radiophosphorus in the nucleated part as compared with the nonnucleated part. Then they presectioned amoebae and put the sectioned portions, nonnucleated and nucleated, into the radiophosphate and demonstrated that there was a rapid uptake of radioactive phosphorus in the nucleated portion and a great reduction of uptake in the nonnucleated portion and, in fact, after a period of about 6 days there was 30 times more radioactive phosphorus in the former portion.

This indicates, therefore, that the phosphate metabolism is profoundly affected in the absence of the nucleus. This result suggests that possibly the synthesis of ATP is affected in the half without the nucleus. However, it was demonstrated by Brachet that the nonnucleated part actually had more ATP and not less. On the other hand, what might be happening here is that the utilization of ATP has been reduced or greatly slowed down in the nonnucleated portion compared with the nucleated portion. This is particularly likely in view of the fact that phosphate uptake is reduced in the absence of a nucleus. This reduction brings us to the very interesting conclusion that we are dealing here with a situation in which the nucleus is responsible for the coupling of respiratory activity with oxidative phosphorylation and it is quite possible, therefore, that when the nucleus is removed respiration is uncoupled from oxidative phosphorylation and this latter process comes to a dead stop even though the respiratory activity is unaffected.

How the nucleus performs this function, whether it is by the production of a coenzyme, as Brachet has suggested, or not is difficult to say. A number of explanations have been but forward of this effect, but one of the most plausible is that, since by far the greater percentage of the DPN synthesizing enzyme is present in the nucleus, its removal deprives the cytoplasm of most of this critical coenzyme. Brachet makes this point too in his discussion of the problem. It was Hogeboom and Schneider who showed in 1952 that the DPN synthesizing enzyme is concentrated largely in the nuclei of liver cells and that the nuclei themselves contain considerable quantities of DPN. It has also been found to be present in fairly high concentrations in the nucleoli of the eggs of starfish. Since the nucleus controls the production of DPN, it is obvious that there would be a drop in DPN in the fraction of cytoplasm in which it is not present.

Now DPN is essential for glycogenolysis and this would explain the holdup in this process in nonnucleated fragments. DPN, however, is also required for respiration, but the DPN concerned with respiration is not present in the cytoplasm but is conserved in the mitochondria and could conceivably not be affected by the process of removal of the nucleus. The DPN which is present in the cytoplasm and which is concerned with glycogenolysis might well be susceptible to breakdown by DPNases in the absence of a nucleus.

The failure of production of more DPN by the nucleus and the extrusion by the former into the cytoplasm would be responsible for this chain of reactions grinding to a stop. Attempts to prove this hypothesis, which was enunciated by Brachet, were made by two authors, one of whom found that the DPN content in the nonnucleated portion of an amoeba decreased by more than half in a period of 75 hrs. after enucleation; the other author reported no change in the DPN level for 6 days. As Brachet says one will need to wait until this conflict is resolved before it is possible to assert or confirm this theory of the relationship of the nucleus to cytoplasmic respiration and oxidative phosphorylation.

When similar experiments are carried out on Acetabularia, the story does not seem quite the same. There appears to be very little difference in uptake of radiophosphorus between Acetabularia with nucleus or without nucleus. However, the situation is complicated by the fact that this alga carries out the process of photosynthesis and that photosynthetic processes are unaffected by the removal of the nucleus and in fact the uptake of radioactive phosphate is concerned with photosynthesis. With Acetabularia it appears then that the nucleus does not exercise the same control as in animal cells.

One of the things that happens when an amoeba is enucleated is that the cytoplasm changes from a fibrillar appearance under the microscope to a granular one. At the same time the basophilia decreases in the nonnucleated portion and continues to decrease for some days after the removal of the nucleus. By the fifth day there is practically no basophilia in the fragment that does not contain the nucleus. All this suggests that there is breakdown of the endoplasmic reticulum, especially the ergastoplasmic part; the loss of basophilia presumably means loss of the RNA particles which were associated with these membranes. Whether this is in fact the proper interpretation of these changes, however, remains to be investigated with the aid of electron microscopy which is only now beginning to be applied to this particular problem.

If this proves to be so, however, we have an example of another control exerted by the nucleus on the cytoplasm, that is, the maintenance of the cytoplasmic fine structure and particularly the content of RNA present in the cytoplasm. We have discussed earlier the possibility that the nucleus may produce RNA and pass it into the cytoplasm of the cell. A very ingenious and interesting experiment was conducted by Goldstein and Plaut in 1955 and described by Brachet. These two authors cultivated organisms known as Tetrahymena which they had immersed in radioactive phosphate solution and the amoebae were then permitted to feed on them. Presumably then the radioactive phosphorus which was liberated into the cytoplasm of the amoebae was absorbed into the nucleus and possibly converted into RNA.

After the digestion of the Tetrahymena, the nucleus from an amoeba which had ingested these radioactive animals was then grafted either into an amoeba which already had a nucleus or into an enucleated amoeba. In the case of the latter, within a relatively short period of time, 12 hrs. or so, radioactive RNA was found to be present in the cytoplasm. That this was RNA was demonstrated by the fact that, after treatment of this organism with ribonuclease, no radioactive particles could be shown by autoradiographic techniques to be present in the cytoplasm.

Another point of interest is that where such a nucleus was poked into an amoeba which already had its own nucleus, radioactive particles of RNA soon appeared in the cytoplasm but these particles did not pass back into the original nucleus which remained free of radioactivity during the whole of the experiment. This indicates that the passage of RNA is probably one-way only, that is, from the nucleus into the cytoplasm.

Brachet has pointed out, however, that Goldstein and Plaut have themselves made the point that the radioactive material which is discharged from the nucleus into the cytoplasm is not necessarily RNA that has been formed in the nucleus. It may, in fact, be a precursor or it may be that the radioactive phosphate is taken up in the nucleus and passed back into the cytoplasm as such and then synthesized as RNA in the cytoplasm itself. However, even if this is so, it is of interest that the nucleus is a source of phosphate which is used for RNA synthesis in the cytoplasm or a source of a precursor which is used for this purpose. It should be mentioned that Goldstein and Plaut made the point that their findings do not suggest that synthesis of RNA in the cytoplasm as such is excluded. All they have, in fact, demonstrated is that RNA or else some precursor or radioactive phosphate is passed from the nucleus into the cytoplasm so that some of the RNA in the cytoplasm may be derived from this nuclear source.

However, in this connection one point which should be remembered is that, in the nonnucleated parts of the amoeba, there is a great drop in RNA following the removal of the nucleus which suggests that a good deal of the cytoplasmic RNA is nuclear in origin. However, it is possible that cytoplasmic synthesis of RNA may be under control of the nucleus by means of a coenzyme, so these experiments are not yet quite conclusive. Brachet makes the point that, if the RNA metabolism is affected in the enucleated amoeba, there should be some inhibition in the synthesis of protein since there are many pieces of evidence which link these two processes together as has been described earlier.

Mazia and Prescott in 1955 found that the amount of methionine labeled with radioactive sulfur (S35) which is incorporated into a nonnucleated Amoeba fragment is two and a half times less than the amount incorporated in the fragment with the nucleus. This difference however does not occur until 3 days after the amoeba has been severed in two. This provides some evidence that there is a link between protein synthesis and RNA synthesis and that in this indirect way, by control over the cytoplasmic RNA, the nucleus also controls the protein synthesis in the cell. It should be noted that protein synthesis does not fall to zero in fragments without a nucleus, so there is a residual cytoplasmic synthesis of proteins which persists even although the nucleus is removed from the cytoplasm.

It is not known whether the nucleus controls the synthesis of all the different types of cytoplasmic proteins. Brachet has studied changes in a number of different enzymes in the nonnucleate halves of amoebae and has demonstrated that various types of enzymes are affected in different ways by the nucleus. (Enzymes are, of course, proteins so the study of the enzymes gives an indication of the synthesis of this particular type of protein.) Protease, enolase and adenosine triphosphatase are enzymes which appear to undergo very little or no change once the nucleus is removed. Amylase, on the other hand, seems to increase slightly in activity then drops back to normal level so that it is not affected very much.

Dipeptidase decreases in the beginning and then remains at a constant lower level whereas acid phosphatase and esterase are reduced progressively and afer a few days cannot be demonstrated as being present at all. Thus different enzymes are obviously under nuclear control to different degrees and it becomes obvious that control of the nucleus over cytoplasmic proteins is a very complex one.

It has been suggested by Brachet that the difference in behaviour of the enzymes might be related to their localization in the cell. He points out that Holter in 1955 demonstrated that amylase and protease were incorporated in very large granules which were either mitochondrial or were possibly the lysosomes of De Duve which we have described earlier. This might suggest, as earlier evidence has possibly demonstrated, that the mitochondria do not appear to be under the control of the nucleus. Dipeptidase is in the cytoplasm proper and is not bound to the mitochondria so one could understand that it might be reduced in the absence of the nucleus. Acid phosphatase and esterase which decrease progressively when the nucleus is removed may possibly be bound to the microsomes (E.R.) as shown by Brachet. Some acid phosphatase seems to be associated with the mitochondria and De Duve has claimed that acid phosphatase is one of the hydrolytic enzymes found in his lysosomes. We have no certainty that all the different types of hydrolytic enzymes are present in the one lysosome; it may be there are different lysosomes, some containing acid phosphatase, some containing cathepsin and others proteases and so on and if this is the case it is possible that some of these are affected by the absence of the nucleus and some are not.

It is also of interest that Danielli and his colleagues in 1955 made hybrid amoebae by putting the nucleus of one species into the cytoplasm of another and then prepared antibodies against the two species. They found that the lysis of the hybrid amoeba is under the control of the nucleus. This appears to demonstrate that the determination of antigenic specific characters, as Brachet puts it, is under the "nuclear dominance while the morphological and physiological character of the hybrid is under cytoplasmic dominancy." It is of interest that in our earlier discussions in this book on the role of the nucleolus on the synthesis of RNA we mentioned that the dissected out nucleoli of Acetabularia exposed to radioactive phosphate had incorporated P32 with extraordinary rapidity into their RNA so this is a very good support of the claim that the nucleolus is a centre of RNA synthesis since it contains nearly all the RNA of the nucleus.

It is of interest that in the case of Acetabularia the removal of the nucleus actually stimulates RNA synthesis in the cytoplasm, but it is also to be noted that in intact Acetabularia the nucleolus is much more active than the cytoplasm in the production of RNA. It has been suggested by Brachet that this can be explained by the fact that the high activity of the nucleolus in production of RNA in intact Acetabularia is due to the fact that the precursors of RNA are snatched away from the cytoplasm before they can be used there for its synthesis. However, the mechanism for synthesis of RNA is present and active in the cytoplasm and the removal of the nucleus (with its contained nucleolus) removes competition for the precursors; the cytoplasm could then obtain all that was necessary.

In the case of "protein synthesis" (absorption of labeled amino acids) in Acetabularia, it has been demonstrated that the amino acids appear in the nucleolus before they appear in the cytoplasm and that once the nucleus is removed, protein synthesis by the nonnucleated portion is actually faster than in the nucleated portion and this "ties in" well with the demonstrated increase of RNA synthesis. It also shows that the nucleus is not essential for the synthesis of protein by the cytoplasm, but further experiments have demonstrated that such anucleate protein synthesis is of a short-term nature (about three weeks) and that for it to be extended over a long period of time the nucleus is necessary. It is of interest that, although the synthesis comes to an end, the turnover (breakdown) of protein continues for quite a number of weeks even if the nucleus is not present.

A striking thing about the intact cell is the metabolic

activity of the RNA contained in the nucleolus in comparison with that found in the cytoplasm. Some early work suggested that incorporation of precursors into RNA was faster in the nucleolus than in the cytoplasm and there is now a good deal of evidence that this is so mostly from autoradiographic studies by Ficq in 1955. By working with the oocytes of echinoderms and amphibians, she demonstrated that radioactive labeled adenine and orotic acid were incorporated very rapidly into RNA in the nucleolus. In the lampbrush chromosomes of amphibian eggs, adenine could be found to be incorporated very rapidly into the loops of the chromosomes, which are known to contain RNA. There have also been some studies with radioactive phosphate and a number of authors have demonstrated that it too is preferentially absorbed into the nucleolus. All these findings, together with the previous literature, strongly support the conception that the nucleolus in particular is a very active site of synthesis of RNA and that it is the source of at least part of the cytoplasmic RNA.

The incorporation of labeled amino acids into the protein of the nucleus has been subject to some small controversy. Originally it was believed that the rate of incorporation into the nucleus was no more rapid than into the cytoplasm, but this was probably due to the fact that the work was carried out on aqueous homogenates and that nuclei lose a certain amount of their protein in such homogenates. It was not until homogenates were studied by nonaqueous techniques that it was found that labeled amino acids were incorporated into nuclear proteins very rapidly.

Brachet points out that the proteins of the nucleus are presumably either the constituents of the chromatin or they represent protein which is localized in the nucleolus and it is not possible to tell for certain with the homogenate technique just where in the nucleus the proteins are localized. Ficq and Brachet have demonstrated that in the liver cells of higher animals the nuclear proteins incorporate radioactive amino acids much more actively than cytoplasmic proteins but in the case of the nuclei of the pancreas, the intestine, the lung, heart, muscle, kidneys, spleen and uterus this is not so. In developing embryos it has been demonstrated that all the nuclei show a very much higher incorporation of radioactive amino acids than the cytoplasm and since, as Brachet points out, there is a considerable synthesis of proteins in the nucleus in actively dividing cells, this is just what would be expected.

The conclusons made by Brachet about this problem is that there is no doubt that the nucleus and particularly the nucleolus has a very active metabolism of RNA but that protein metabolism is not more active in the nucleus than it is in the cytoplasm, the exception being those cells which are undergoing rapid division. This is of interest in view of the fact that, since RNA is being made relatively more rapidly than proteins are being synthesized in the nucleus particularly in the nucleolus, more RNA is being formed than is necessary for this purpose. It seems likely, therefore, that the excess RNA passes to the cytoplasm.

Brachet has given a general discussion of the possible relationships of the nucleus and the cytoplasm and the readers are referred to his book for full details. However, certain of the more essential conclusions might be repeated here. One of the functions which the nucleus could perform since it is relatively poor in enzymes is the production of coenzymes and it is possible that some of the enzymes found in the cytoplasm might be under the regulation of the coenzymes secreted by the nucleus. The importance of coenzymes might explain the association of the mitochondria with the nuclear membrane which has been recorded by quite a number of authors. However, we only know for certain that DPN as a coenzyme is made by the nucleus and we do not know for certain what other enzymes might be produced by it. It is of interest that Tewari & Bourne have produced considerable evidence that the nucleolus in spinal ganglion neurones synthesize ATPase and glucose-6phosphatase and passes it first to the body of the nucleus then to the cytoplasm.

The striking effect, however, of this association is that the activities of the mitochondria seem to be pretty well independent of the nucleus in so far as cellular oxidation is concerned since the oxidation proceeds whether the nucleus is there or not. However, as we have pointed out earlier, it is an interesting possibility that the nucleus acts as a device for coupling by some means or other oxidative phosphorylation with respiration. The independence or otherwise of the mitochondria from the nucleus is of interest and Brachet points out as we have done earlier in this discussion that the nucleus may in fact depend upon its supply of energy (ATP) from the mitochondria.

Where, however, the nucleus appears to exert its effect most strongly is on the endoplasmic reticulum and, since it has been (demonstrated that great decrease of basophilia and change in the nature of the cytoplasm occurs when the nucleus is removed from the Amoeba, it seems possible that the real relationship between the nucleus and the cell lies in the control by the former over the endoplasmic reticulum. Its control here may be dependent on the secretion of RNA granules which become attached to the reticulum.

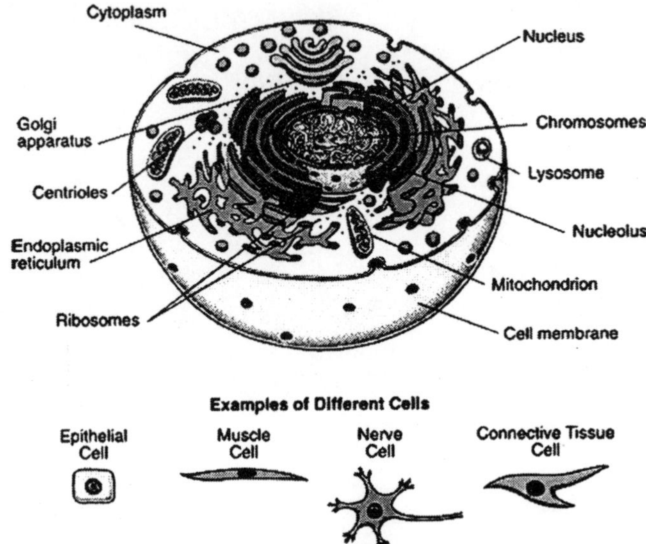

Fig. Division of Labour in Cells

Originally it was suggested by Caspersson that DNA was really the nucleic acid fundamentally concerned with the synthesis of complex proteins. More recently it has been suggested that DNA probably directly synthesizes the proteins of the chromosomes, in other words it causes a reproduction of the protein portion of the genes. It is very likely that RNA plays an important part in embryological processes and it might be of interest in concluding this chapter to repeat Brachet's conception of its role.

"Since the synthesis of proteins is primarily a cytoplasmic process and since, as we know, the RNA present in the ergastroplasmic small granules certainly plays a part in the synthesis it is tempting to visualize the whole process in the following, entirely hypothetical way. Under the influence of the nuclear genes, specific cytoplasmic RNA's would be synthesized: these RNA's are probably not entirely synthesized in the nucleus and it is possible that interactions between purely cytoplasmic RNA and RNA of nuclear origin occur. The specific RNA's would in turn organize the synthesis of specific cytoplasmic proteins according to the well-known template hypothesis."

Chapter 7

The Specialized Cells

We have been dealing with cells in a rather general way, using a variety of cell types to illustrate points concerning cell physiology. It is proposed now in the last stages of this book to consider two or three specific types of cells. To begin this chapter we will consider the glandular cell and, since there are a wide variety of both exocrine and endocrine glands in the body, comments will be restricted to the type of secretion process which is demonstrated particularly well either in the salivary glands or the pancreas and particularly we will refer to the pancreas, although some details of the salivary glands will be included as well.

The mechanism of secretion in externally secreting glands has been a subject of study for a good many years and very detailed work was published on this subject by Nassonow in 1923; Bowen in a series of publications between 1924 and 1929 fully confirmed and extended Nassonow's work. The general trend of studies by Nassonow and Bowen was the relationship of the Golgi material in particular to the production of secretion droplets. They both demonstrated that the first droplets of secretion which are visible under the optical microscope appeared in the interstices of the Golgi network, that as the number and size of these droplets increased, the Golgi network became hypertrophied and that when the cell carried its full load of secretion the apparatus appeared to break up and small pieces of it appeared to become attached to some of the secretory granules.

Studies have been made on the pancreatic cells of external secretion (as distinct from the islets of Langerhans) from the point of view of fine structure. The pancreatic cell structure can thus be taken as a general model for the cells of external secreting glands.

Two groups have made a special study of the fine structure of the acinar cells of the pancreas. These two are Ekholm and Edlund who investigated the human pancreas and Palade and Siekewitz who studied the guinea pig pancreas. Generally speaking there was little difference between the two types of pancreas. The same cytoplasmic elements--the endoplasmic reticulum with associated ribonucleoprotein granules, the mitochondria and the zymogen granules were present in both.

In guinea pigs which had been starved, Siekewitz and Palade demonstrated that there was a well-formed endoplasmic reticulum (ergastoplasm) and that the cisternae (spaces between the pairs of membranes) were very small and had special orientation within the cell. There were very few granules in the cisternal spaces. The apical parts of the cell contained a number of zymogen granules and the lumina of the acini were very small and appeared to be occupied by an amorphous material which had a density similar to that of the zymogen granules. This finding agrees with the statement that secretion of zymogen does occur from the pancreas in small amounts even in long starvation. In guinea pigs that had been killed one hr. after feeding, there were many granules within the cisternae of the endoplasmic reticulum which were greatly distended. The number of zymogen granules found in the apex of the cell was generally less than in the starved animals. The lumina of the acini were distended and contained irregular and illdefined masses of electron dense material which was probably discharged zymogen.

Siekewitz and Palade subsequently made a very interesting study of the functional variations in the microsomal fractions in the pancreas of the guinea pig. The microsomal fraction is mentioned before represents the endoplasmic reticulum plus its attached ribonucleic acid particles. Their experiments started from the fact that they observed that, when the microsomal fraction from starved guinea pigs was analyzed for enzyme activity, it was found that the ribonuclease and TAPase (trypsin activatable protease) activities varied a great deal in different preparations and in some cases reached as high as 30% of the activities found in the zymogen fraction.

The zymogen fraction is that part of the homogenate of the pancreatic cells which is formed from the granules in the cell that are described as the zymogen granules and which are primarily comprised of digestive enzymes and precursors of enzymes-hence the name "zymogen." Since the zymogen fraction is composed almost exclusively of zymogen granules and since these have a higher total RNAase and TAPase activity than any of the other cell fractions, the authors wondered whether this similarity indicated some functional connection and they initiated a series of experiments in order to test this out.

In other investigations they had found that in some cases dense granules had been found within the cisternae of the endoplasmic reticulum in electron microscope preparations of the exocrine cells of the pancreas of the guinea pig. These intracisternal granules could be compared, both from the point of view of fine texture which they demonstrated and their density, with the zymogen granules. However, there was a considerable difference in size and the zymogen granules appeared to be outside the endoplasmic reticulum whereas the smaller particles were actually in the cisternae between the double membranes. Further studies have demonstrated that these intracisternal granules are present in large numbers in this position about 1 and 2 hrs. after starved guinea pigs have been fed but that by 3 to 4 hrs. they have more or less disappeared.

The authors consider that, since it is the endoplasmic reticulum which forms the microsomes in homogenization, perhaps some of the high enzyme activity demonstrated in the microsomal fraction might have been due to the enclosure of some of these granules which were present within the microsomal cavity. They investigated this problem by taking various fractions from starved and fed animals, collecting the glands about 1 hr. after feeling, the time when they expected to obtain the highest concentration of granules within the cisternae of the endoplasmic reticulum. They were able to obtain subfractions an hour after feeding which consisted largely of cisternal granules together with some detritus derived from the microsomal fraction. They found that the TAPase and the RNAase activities present in this granular fraction were actually higher than those of the microsomes from which they were presumably derived and were sometimes as high as the values for the zymogen fraction as such. A similar increase in activity was also described for the enzyme amylase.

The granules present in the cisternae were already in a finished state and were probably destined therefore to become zymogen granules, but their precise position in the secretory circle is not clearly seen at the moment. It is known that in the late stages of the secretory cycle of the pancreas the large vacuoles in the Golgi zone become filled up with a material which has a high electron density and it is possible that these intiracisternal granules feed into the Golgi material and there condense into large granules which are passed to the apical part of the cell. Hirsch has said that in many cases it appears that the granules which are shed into the lumen of the pancreas appear to be enclosed in a structure resembling a membrane of cytoplasmic material. This might be explained in the following way.

Assuming that there is a continuous passage from the endoplasmic reticulum lumen through the cisternae in the Golgi apparatus and so to the exterior, it is possible that Siekewitz and Palade's intracisternal granules pass to the Golgi apparatus and there coalesce to form large granules which then pass away from the Golgi material toward the apical part of the cell along those parts of the endoplasmic reticulum which lead to the exterior of the cell. Since these granules are very big when the leave the Golgi apparatus, they no doubt cause great dilation of the cisternae and it is even possible that as they are pushed out of the cell they tear part of the double membranes from the ergastoplasm and carry this as an investment as they pass out into the exterior. Even if there is no continuous passage it is quite possible that the droplets might be excreted from the cell covered with a submicroscopic skin derived from the endoplasmic reticulum or the Golgi membranes or simply composed of protein secreted on to them by one of these organelles.

Siekewitz and Palade point out that in their preparations, although they found a very high proteolytic RNAase activity in the isolated intracisternal granules, quite considerable proteolytic activity and RNAase activity were also found in the less dense fragments of the microsomes which presumably are composed simply of endoplasmic reticulum and the small RNA particles attached to them. It appears very likely that this activity is localized in the RNA protein particles. It is possible that these individual particles sypthesize enzymes, become greatly enlarged and pass through the membrane of the endoplasmic reticulum into the cisternae.

This represents a scheme for the production of secretion granules by the cell and the joint role of the endoplasmic reticulum, the mitochondria and the Golgi apparatus in this process. According to this scheme, in the pancreas amino acids become associated with the RNA granules of the endoplasmic reticulum membrane (left of diagram) and are passed as preprotein (polypeptides) into the cisternae where they grow in size to become the intracisternal granules of Palade and Siekewitz this stage is headed "1 hr. after stimulation" in the diagram).

In the text of this book the previous observations of Hirsch, in which he has seen "secretion" droplets in the pancreatic acinar cell moving through the cytoplasm to the Golgi zone have been mentioned, these droplets are approximately the same size as the intracisternal granules described by Palade and Siekewitz. Hirsch now believes that his moving droplets represent the intracisternal granules passing along the endoplasmic reticulum. The large secretion droplets of gland cells appear in the Golgi apparatus and if the intracisternal granules represent presecretory material their transference from the endoplasmic reticulum to the Golgi material requires some explanation. There is no evidence that the cisternae of the endoplasmic reticulum are continuous with the Golgi material but this possibility must be considered.

Hirsch believes that the mitochondria provide the energy required to dissolve the cisternal granules, to pass them through the membranes of the endoplasmic reticulum and through the Golgi membranes inside which they become concentrated into droplets of secretion, growing larger and larger until they are converted into zymogen granules, surrounded by a membrane and pushed to apex of the cell ready for discharge. This is the morphological story to date as told by the electron and light microscopes.

The metabolic story has been studied, using salivary gland material by Junquiera and his colleagues with rats as their experimental animals. Normal cats and rats with ligated salivary ducts were used. The significance of ligating the excretory duct is that it leads to cessation of secretion, disappearance of granules of secretion, decrease in size of the cells and the gland as a whole and there is a corresponding decrease in activities of the enzymes. The gland does not undergo any degenerative process. Junquiera and his colleagues used these types of glands to study not only the histochemical and biochemical changes in the cells but they tried also to study the morphological changes as demonstrated by the light microscope in an attempt to equate the various activities and morphological elements in the cell with the processes of secretion.

These results are given in summary as follows. In the control gland it was found that there were many mitochondria, they were mainly in the form of short rods through some round specimens were present. As a result of ligation they were very greatly reduced in numbers. Millon's histochemical test for proteins demonstrated a strong reaction in the normally secreting gland and only a very weak reaction in the gland which had been ligated. Studies in basophilia (to demonstrate degree of accumulation of RNA) showed a very strong basophilia in the control gland, but in the ligated gland which was apparently not secreting at all the basophilia was very weak. Biochemical investigation of the ratio of ribonucleoprotein over deoxyribonucleoprotein (RNAP/DNAP) showed this ratio to be 2.18 in cells of the control gland but only 0.84 in the ligated gland, so there was obviously a very profound decrease of RNA by comparison with DNA in the nonsecreting gland.

Biochemical studies of protease activity demonstrated that in the control gland it was 560.0g. phenol/100 mg. tissue and in the ligated gland it was 266.0 ?g. phenol/100 mg. tissue. In the mouse, the cathepsin activity was 16.9g. tyrosine/1g. DNA protein and in the ligated gland it was not reduced quite as much as some of the other enzymes but was down to 10.3 ?g. tyrosine/1g. DNAP. Alkaline phosphatase varied a little between rats and mice. In rats the figure was 2.8 mg. phenol/ 100 mg. tissue and in the ligated gland it was 1.7 mg. phenol/ 100 mg. tissue so this was a reduction of something like a third. In mice curiously enough, in the control gland there was 2.0 mg. phenol/100 mg. tissue of alkaline phosphatase and in the ligated gland it actually went up to 2.4 mg. phenol/100 mg. tissue. So it seems that if the alkaline phosphatase is playing a significant part in the secretory processes of rats, it is a specific process and it does not play the same part in mice.

The acid phosphatase results were interesting because in rats the control gland showed 1.8 mg. phenol/100 mg. tissue activity and in the ligated glands there was no change at all. In mice, however, in the control gland there was 2.0 mg. phenol/100 mg. tissue activity and in the ligated gland 0.42 mg. phenol/100 mg. tissue, so that, perhaps, if phosphatase is playing a part in the secretory cycle, then in the rat it is alkaline phosphatase that is involved and acid phosphatase in the mouse.

Succinic dehydrogenase was studied by Thunberg's methylene blue method and there was found to be high activity (50% reduction in 12 min. of methylene blue) in the control gland and in the ligated gland there was a 12% reduction in colour, which was decreased in mice to less than a quarter. There was not quite as much reduction in rats; using the Warburg apparatus there was a 63% increase of the QO2 after the addition of succinate and in the ligated glands only 28% increase. So there was a significant reduction in oxidative activities which was confirmed by the cytochemical indophenol oxidase test for cytochrome oxidase, the control gland cells giving a dark blue and ligated glands only a light blue colour with this reagent. The oxygen consumption in the rat cells gave a QO2 of 4.2 in the control gland and only 1.8 in the ligated gland. The figure for glycolysis and the glycolytic quotient in relation to nitrogen was 3.2 in the control gland and 3.3 in the ligated gland, so there was significant different in the glycolytic rate. Pyruvate utilization similarly dropped by about two-thirds and thus it appears from these figures that not only is there a drop in respiratory activity of the cells when they are not secreting but there is also a drop, as one might expect, in oxidative phosphorylation.

It is of interest, however, that glycolysis is maintained at a constant rate in both ligated and nonligated cells and, as Junquiera and his colleagues point out, it appears that the energy derived from glycolysis itself is used by the cell largely for its basic needs and not for its specialized function as a secreting cell. It appears that the ATP, ADP and phosphocreatine are really the compounds that are the sources of energy for the secretory process in the cell since there is such a spectacular decrease in these following ligation. It is of interest that the activity of cathepsin in the secretion of the salivary glands seems to be controlled in vivo by the male sex hormones. There is a parallel between cathepsin activity and protein synthesis and the drop in activity of cathepsin in the cells of ligated glands may be significant from this point of view.

Junquiera and Hirsch have also described the light-microscope changes that take place in cells of ligated glands. The Golgi apparatus is undoubtedly related to the processes of secretion if the early work which we have mentioned before can be taken at its face value. Junquiera and Hirsch summarized the evidence for the participation of Golgi material in the process of secretion as follows. The Golgi apparatus changes in size and structure very considerably during the secretory cycle. Droplets appear in the substance of the Golgi material and gradually appear to be transformed into secretory granules. This fact has been observed not only in fixed and stained preparations but also in the living pancreas, for instance, by Hirsch in 1932.

In 1939 he suggested that the Golgi bodies are the sites in the cell where cytoplasmic products congregate and become formed into zymogen granules. There is no biochemical data, as has been pointed out both by the present author and by Hirsch, that the Golgi bodies are actually the site of protein synthesis since they contain no RNA and do not appear to contain oxidative enzymes or other enzymes concerned with energy production, although in some cells such as the absorptive cells of the gut they do contain phosphatase, ATPase (heat stable), alkaline phosphatase, acid phosphatase and so on. In 1954, Sjöstrand and Hanson demonstrated by ultrastructural studies that there was an intimate relationship between the Golgi vacuoles, the zymogen granules, the ground substance of the Golgi apparatus and the alpha cytomembranes (endoplasmic reticulum): "there are all transition stages observed between, on the one hand, bodies with a pronounced elongated form and with the most direct topographic relations to the Golgi membranes and Golgi granules embedded in the ground substance on the other.

The impression when observing these pictures is that they show snapshots of the conversion of membranes into granules and vice versa. The small granules seem to coalesce to bigger granules which gradually gain the size, form and opacity of the zymogen granules."

This work appears to confirm the conception of the Golgi material as a condensing or aggregating region of the cell and reminds one of the statement of Kirkman and Severinghaus in 1938. "A great deal of work strongly suggests that the Golgi apparatus neither synthesizes secretory substances nor is transformed directly into them but acts as a condensation membrane for the concentration, into droplets or granules, of products elaborated elsewhere and diffused into the cytoplasm. These elaborated products may be lipoids, yolk, bile constituents, enzymes, hormones or almost any other form of substance." We do not know, of course, to what extent these synthetic processes are controlled by the mitochondria. There are papers in the literature suggesting that the mitochondria actually produce the zymogen granules in the pancreas and it seems very likely that in parts of the endoplasmic reticulum the mitochondria come in very intimate contact with the nucleoprotein granules on the outside of the membrane and that they supply the energy for the synthetic processes taking place in these particles. At any rate the particles of enzyme protein that are produced pass into the spaces and presumably undergo some benefit from contact with the Golgi material and possibly even from contact with the nucleus.

One could conceive of them as all passing into the cisternal space which surrounds the nucleus, but to what extent the metabolic activity of the nucleus might affect any enzyme protein passing through this space one cannot say for certain. Perhaps the passage of the granules to the Golgi region serves not only as a mechanism whereby the granules are aggregated to a large zymogen granule but possibly something is done to prepare the enzymes for activity upon secretion. Also one should consider the possibility that vitamin C, despite the known defects of localization in the technique, may occur, as the present author has claimed, in the Golgi apparatus at times of great synthetic activity and there is a possibility that in passing through into an area which is saturated with a reducing substance, oxidation of the contents of these granules prior to their excretion may be prevented.

So we can see a division of labour very clearly in the process of secretion in the gland cell but the interesting thing, as we have been emphasizing right through this book, is that we are not dealing with an isolated division of labour but an interdigitating division so that each job fits in very well with the job done by another part of the cell.

THE STRIATED MUSCLE FIBRE

The muscle fibre is composed of four fundamental constituents:

- The sarcolemma which is the equivalent of the cell membrane;
- The fibrils which represent the structural elements responsible for contraction;
- The sarcosomes which are really the mitochondria and which contribute the supply of energy for muscular contraction;
- The sarcoplasm which is the ground cytoplasmic substance in which the other structures of the muscle fibre are embedded.

In addition, there is a specialized region of the fibre called the motor end-plate at the point where the motor nerve fibre makes junction with the sarcolemmal membrane and which is responsible for initiating the contraction of the fibre. For the moment we will discuss very briefly the structure of each of these elements and we will include the description of the structure of the caveolae introcellulares with an account of the structure of the sarcolemma.

Muscle fibers themselves have three main shapes. They may be cylindrical with ends which are conical, they may be spindle-shaped with extremities tapering off very finely, or they may be conical with one end long and attenuated and a broad base at the other. This depends, of course, on their position in the muscle itself. The first of these types of fibers usually runs the whole length of the muscle, the second type is usually situated within the main part or belly of the muscle, the third type is usually attached to a tendon at one end and the other end terminating somewhere in the interior of the muscle itself. The shape of a transverse section of muscle is oval or spherical when it is fresh but it shrinks considerably following fixation and one usually finds rather angular cross sections of the fibres. Voluntary muscle fibers vary from about 10-100? in diameter. Presumably this difference in size is associated in some way with the amount of work which the muscle has to perform. There are in any single muscle, fibers of many different diameters. Fibers vary very much in length and may in some cases extend only a few millimeters and in others, fibers in excess of 34 cm. in length have been seen.

Schwann in 1839 and Bowman in 1840 described a thin membrane which surrounded the voluntary muscle fibre. It was described by Schwann as the cell membrane of the fibre and this was later agreed to by Bowman. It can be demonstrated by placing a muscle fibre in fresh water or by causing a sudden coagulation of the contents which then contract. The membrane or sarcolemma, as demonstrated in this way, seems to be a pale, colourless and apparently, under the optical microscope, structureless membrane and it is semipermeable since if placed in water it swells by imbibition and if subsequently placed in concentrated sugar solution this water is removed. It was thought to be about 0.1 across and to have a slight infolding at each Z band.

Others have shown that the sarcolemma is more complicated, i.e., under the electron microscope there are two dark osmiophilic lines separated by a light osmiophilic line which together are approximately 300 A across, each of the lines in this triple structure being about 100 A thick; also there is a thick layer of dense material about 500 A thick which extends inward from the inner of the dark lines of the membrane.

Under the light microscope the early workers had described the sarcolemmal membrane as being about 0.1? thick which is in fact 1000 A. We will see that, if we add the 500-A thick complex to the three lines which themselves total about 300 A, we get a figure of about 800 A which is pretty close to the 0.1? size of the sarcolemma described by the optical microscopists. As Robertson points out we are faced, of course, with the decision as to which of these many structures can be regarded as the true membrane. What we are in fact dealing with here is really a membrane complex rather than a single membrane and this, strictly speaking, applies to the membranes of all other cells as well. Every now and again the sarcolemma is penetrated by holes which are the ends of the caveolae intracellulares.

Some of these little openings or caveolae expand into vesicles and in some cases both the vesicles and the caveolae themselves may extend in a complicatd fashion, branching and ramifying over and in between the myofibrils. It has been suggested by Bennett and also by Palade that such caveolae are not a standardized feature of the structure of the sarcolemma of the muscle fibre, in general, but that they represent a mechanism whereby ions or macromolecules such as carbohydrates or proteins can obtain access into the interior of the fibers without violating the osmotic properties of the membrane.

According to the nature of the binding sites there, the membrane would then infold, form a caveola and this would seal off to form a vesicle and the vesicle would then pass into the interior and might then be destroyed so that the substances become free in the cytoplasm. This is a pinocytotic type of interpretation of these caveolae and the possibility exists, of course, that this type of activity may occur in many types of cells. Compounds could be passed out of the muscle fibre in a similar way, that is, by becoming attached to the internal surface of the plasma membrane which would then surround it and open up to form a caveola and the contents would become liberated to the exterior. Thus large macromolecules could get through the membrane without actually piercing it.

It is of interest that in connection with this theory Whatman and Mostyn noted droplets of fat which had been stained with fat dyes in the process of passing through the sarcolemma of cardiac muscle fibers. Quite a number of cells and perhaps all cells appear to have attached on the outside some sort of polysaccharide coating to the plasma membrane. Such a coating occurs on the surface of mammalian erythrocytes where it is extremely thin and it has been pointed out that this may represent the surface coating which causes the ABO agglutination reaction.

Plant cells are, of course, expert in putting polysaccharides around the outside of their cell membranes and bacteria consistently do this too. Outside the sarcolemma there also appears to be a cloud of material, approximately 500 A thick which is possibly polysaccharide material. It is of interest that there is a positive periodic acid-Schiff (PAS) reaction in the sarcolemma and this may be the cause of the reaction. This significance of the polysaccharide complex may be its ability to bind various molecules which are to be taken into the interior of the muscle membrane.

Fig. Cholinesterase in Human Neuromuscular Junctions

In the region of the motor end-plates there is considerable modification of the sarcolemma and Robertson in 1956 and Couteaux in a series of papers have published detailed accounts of this. The sarcolemma in the region of the motor end-plate is very much thicker, denser and more folded particularly into a series of troughs which penetrate deeply into the sarcoplasm of the muscle fibre. The external polysaccharide-like material just described is present everywhere and separates the actual endings of the neural components of the motor end-plate from the sarcolemmal membrane itself. The sarcoplasm of the muscle fibre was originally described by Spidel in 1939 as being in a gelled condition, this can change to a sol if there is any damage to the muscle.

The sarcoplasm not only surrounds the nucleus and is obvious in that part of the fibre but it also penetrates and surrounds and fills in the spaces between the myofibrils themselves and between the myofibrils and sarcolemma. The relative amounts of sarcoplasm in the fibers varies very considerably and it is of interest that Golarz and Bourne have demonstrated great activity for the enzyme acetyl phosphatase in the sarcoplasm of human muscle. The sarcoplasm also contains most of the soluble proteins of the muscle fibre. Under the electron microscope it is less dense than the other components of the fibre. There are also a number of small granules scattered about the sarcoplasm which show a great range in size, something between ten and a couple of hundred angstroms across. They are distributed without any particular pattern and in some regions of the sarcoplasm there do not appear to be any at all. They appear to be more abundant in certain areas of the sarcoplasm in heart muscle fibers and, according to Fawcett and Selby they may, in fact, be composed of glycogen. Such areas give a strong PAS reaction and, if the section is treated with amylase prior to application of the PAS reaction, the reaction disappears. This is strong evidence that these granules are, in fact, glycogen.

Beckett and Bourne have demonstrated that PAS positive material was present which was dispersed irregularly in human muscle fibers, sometimes as irregularly distributed aggregates and sometimes as fine granules. To some extent the difference in size of these granules depends on the changes which take place following death. A biopsy specimen, for instance, shows only small granules and aggregations but by 12 hrs. the aggregates have reached their optimum size and after 48 hrs. autolytic procedures have removed them altogether. Beckett and Bourne have demonstrated that the amount of stainable material with the PAS technique in normal muscles is extremely variable and that there is no correlation between the amount present and the anatomical site of the muscle concerned. Also in muscular and neuromuscular disorders the amount of stainable material is variable and inconstant and shows no particular relationship with the neuromuscular diseases which were studied.

It is of interest that Dempsey and his colleagues and Beckett and Bourne were unable to remove all this PAS positive material with either diastase or saliva. This suggests that we are dealing either with glycogen which is linked to and protected by protein in some way or with a mucopolysaccharide type of substance or possibly even aldehyde groups are being produced by the PAS reaction by lipids or fat. Beckett and Bourne also applied the McManus PAS technique to formalin fixed, frozen sections in conjunction with the Sudan black test for fats. These were made both on rat and human muscle. By this frozen section method it was found that the muscle fibre was stained in a regular diffuse positive PAS background staining and there were also droplets of the material giving a strikingly intense positive PAS reaction. These droplets are often considerably increased in pathological muscles and appear to accumulate at points of mechanical damage in muscle fibers. These strongly positive droplets do not appear to be glycogen. Paraffin embedding of the muscle, however, removes these droplets. It is possible because of these facts that they may be a lipid of some sort. They are not, however, sudanophilic (thus are probably not fat) and do not give a Schultz reaction for cholesterol.

Many muscles fibers contain droplets of fat in the sarcoplasm. These vary very considerably in individual muscle fibers, in individual muscles and in individual muscles as a whole. Some muscles have more fat than other muscles. This was first described as long ago as 1841 by Henle and subsequently investigated by Kölliker and others. Fat was recorded in human muscle as long ago is 1889. Beckett and Bourne using the Sudan black technique in human muscle found that fat could occur in droplets of very variable size, distributed at random, but that a few were situated at the poles of the nuclei. Also there was some staining in the regions of the cross striations with Sudan black.

Some muscle fibers have a red colour and some have a white colour. Red and white muscles are well known amongst mammals and in animals such as the rabbit and guinea pig and also in the turkey both red and white muscles occur separately, but generally speaking, particularly in humans, muscles are a mixture of both red and white fibers. The red colour of the fibre depends upon the fact that it contains myoglobin or muscle hemoglobin. The red fibers contract more slowly and remain contracted for a longer period than the white muscles. There are more of them in those muscles which are concerned with posture. Generally speaking too, the red muscle fibers contain more sarcoplasm than the white and they also appear to contain more fat which presumably functions as a reserve supply of energy for the muscle fibre.

Ribonucleic acid granules have not been identified in the sarcoplasm of adult muscle. An interesting new development in the structure of muscle fibers is the delineation of the sarcoplasmic reticulum described by Bennett in his chapter on "Structure and Function of Muscle." As Bennett points out, for more than 100 years papers have reported that there appears to be some sort of a network or series of tubules associated with muscle fibers. Retzius in 1881, had described a sarcoplasmic network and Kölliker had referred to a network in 1888, however, the most detailed treatment was given by Veratty in 1902. He described a "reticular apparatus" in muscle, this is a confusing term since the Golgi apparatus had often been referred to in these terms. Bennett and Porter in 1953, found this component in electron-microscope pictures of muscle and described it as a sarcoplasmic reticulum.

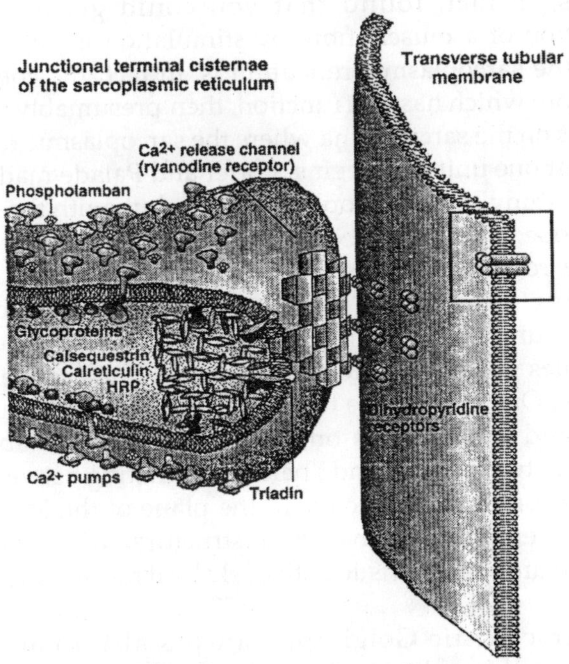

Fig. Sarcoplasmic Reticulum

Sjöstrand and his colleagues subsequently referred to them as sarcotubules. It is of interest that the work on this reticulum which was described and understood very well by these early workers appeared to be completely ignored in later books of histology where one finds scarcely any reference at all to what now appears to be a very important structure in the muscle fibers.

The work of Bennett and Porter and a number of other investigators has demonstrated that this reticulum really resembles in its fine structure the endoplasmic reticulum found in other cells. Its membranes have the triple structure comparable with what Robertson described as the "unit membrane." These tubules appear to extend inward from the sarcolemma. The work by A. F. Huxley and his colleagues has indicated that there is probably a structure in the muscle fibre which is capable of transmitting depolarization of an ion aligned membrane into the interior of the muscle. These authors, in fact, found that you could get a localized contraction of a muscle fibre by stimulation in certain spots and, if the sarcoplasmic reticulum is, in fact, the component in the fibre which has this function, then presumably these are the spots on the sarcolemma where the sarcoplasmic reticulum or at least one unit of it begins. Porter and Palade made a very detailed study of the sarcoplasmic reticulum with the electron microscope.

The reticulum appears to be arranged in the form of alternating sets of tubes which anastomose with each other and surround the fibre like a bracelet. The anastomosing of the tubules gives the general impression of a lacelike type of structure. One set of these tubules has a plane of symmetry at the Z band and extends on teristic both of it to about the junction between the A and I bands, alternating between these is a series of tubules which is in the plane of the M bands. It should be emphasized that these structures which are being described are characteristic both of skeletal muscle and cardiac muscle.

Characteristic Golgi apparatus is also found in the sarcoplasm of muscle close to the nucleus as one might expect.

Although we have referred to a striated muscle fibre as a cell, probably it is really a number of cells since any one striated muscle fibre may contain a number of nuclei and in a very long fibre, which measures several centimeters in length, some hundreds of nuclei may be present. They lie at the surface of the fibers (except in cardiac muscle where they are in the centre) just under the sarcolemma surrounded by sarcoplasm. The nuclei are generally oval in shape, with their long axes parallel with the long axis of the fibre, they are approximately 8-10? long and may be extend as much as 17?. They are hard to see in unstained fresh muscle fibers but come out very well in fixed and stained preparations. In muscle diseases and in degenerating muscle, they appear to migrate into the centre of the fibre and one can see long strings of muscle nuclei aligned end-to-end along the centre of the fibre. It is of interest that in embryonic muscle fibers the nuclei occupy a middle rather than a hypolemmal (close to the sarcolemma) position.

The fine structure of muscle nuclei does not differ from those of other cells. The nuclei are surrounded by two unit membranes analogous to the membranes of the endoplasmic reticulum of other cells. In fact, Bennett thought the outer membrane of these pairs may be continuous with the cytomembranes which are found in the perinuclear cytoplasm or even continuous with those of the sarcoplasmic structure is fundamentally the same in muscle fibers is in other cells. The membrane of the muscle nucleus also contains a number of pores.

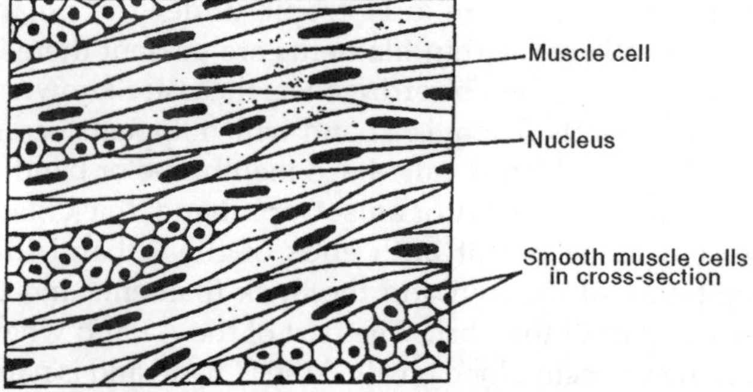

Fig. Cross Striations in Skeletal Muscle Fibers

The fibrils of the muscle fibers are the contractile elements. They appear to extend for the entire length of the fibre and they are approximately 1-3? in diameter. Muscle fibers under the light microscope show a well-determined cross striation and this, of course, is visible in the fibrils and it is of interest that the individual fibrils are aligned so that the appropriate segments of their bands coincide with each other. Between the fibrils are the sarcosomes or muscle mitochondria, to which we will return in a little while.

It is interesting to note that the fibrils consist simply of orientated protein molecules and do not have a surface membrane, therefore the substances in solution in the cytoplasm can have a direct influence on the protein of the fibrils. They have unrestricted access to each other.

The transverse striations of the fibrils of the muscle fibre are due to a difference in density. These cross striations form a pattern and each pattern repeats along the whole of the length of the fibre and each repeating part of the pattern is known as a "sarcomere." In vertebrate muscle each sarcomere is about 2-3 in length. The fibril itself is constructed of filaments which are made of protein and which lie parallel to the long axis of the fibre and overlap each other. There are two kinds of these filaments, one being about twice the diameter of the other and they are also different in length. These two types alternate along the length of the fibril and they overlap at parts and interdigitate with each other. The regions where the thick filaments are located are known as the "A" bands, these are extremely dense and are anisotropic under the polarizing microscope. Where the thin filaments are present there is less density and there is less birefringence, these are known as the I or isotropic bands. There is a band which appears across the middle of these which is called a Z membrane or band.

At the centre of the A or anisotropic band, thick filaments are present alone but at both ends of the A band there is interdigitation of the thin and the thick filaments and this is the densest part of the fibril. The part of the A band where the thick filaments occur alone is called the H band and is naturally not as dense as the rest of the A band. The Z membrane is due to the presence of a band of amorphous material which occupies the spaces between the filaments at the mid-point of the area where the thin filaments are located (the I band). In the case of the A bands, thickening of the thick filaments themselves about the middle of the band causes them apparently to be cut across by what is described as the M line or as the M strip by some authors.

The filaments which compose the fibrils of the muscle fibre are made up of protein and there appear to be three proteins concerned: myosin, actin and tropomyosin. There is more myosin than any other protein; it constitutes approximately 54% of the total protein of the fibrils. There is only about 11% of tropomysin but about 20-25% of actin. This leaves a deficiency about 10% in total protein and it is possible, according to Huxley and Hanson in their chapter in the "Structure and Function of Muscle," that the figures for the various proteins are a little too low particularly for tropomyosin and probably for actin too.

It appears that myosin is the protein of the thick filaments which constitute the A bands of the fibers. Actin, on the other hand, appears to be the protein which makes up the thin filaments of the I bands. Tropomyosin also appears to be present where the actin is situated. In vitro actin and myosin form a complex called actomyosin and this can be persuaded to contract under the influence of ATP just as the muscle fibre contracts under appropriate stimulation in vivo. Each filament of myosin appears to contain 425 molecules. If myosin is treated with trypsin, it splits into two types of meromyosins--one which is called light and the other heavy meromyosin. These meromyosins apparently occur as distinct units in the myosin molecule. Huxley and Hanson suggest that the backbone of a filament of myosin is made up of light meromyosin units which are longitudinally arranged and staggered and the heavy meromyosin units, which are attached to the light meromyosins and project from the filaments, these projections can be seen under the electron microscope.

The number of molecules in each filament of actin can also be calculated and they amount to approximately 600. There is some evidence that the actin filaments contain tropomyosin and, according to Huxley and Hanson, there are about 1.7 molecules of actin to 1 of tropomyosin. The way these two proteins are linked in the filaments is not known.

Huxley and Hanson believe that the actin molecules are exposed at the surface of the filament so that they may form actomyosin links with the neighboring myosin filaments where they interdigitate. The difficult thing to explain is the fact that a muscle can contract very strongly without significant change in the length of its filaments. There are a number of experiments which have been carried out to elucidate this point. However, it appears that what probably happens is that the shortening of the muscle fibre takes place by a sliding of the actin filaments in between the myosin filaments of the A band; they already interdigitate at their ends and the actin filaments simply slide further in.

In the noncontracting muscle the little bumps which have been mentioned before (projections on the heavy meromyosin filaments) come in contact with the actin filaments at the regions of interdigitation and these lock to form an actomyosin complex. Now this happens when no ATP is present, but when ATP is present these locks break apart and the actin filaments, can slide up into the myosin filaments.

When ATP is dephosphorylated then the locks will form again and it is possible that in contraction the actin filaments slip into the myosin filaments point-by-point, there being a locking and releasing corresponding to the production and dephosphorylation of ATP.

Then when all the ATP formed is dephosphorylated the bands lock again. As soon as ATP is re-formed then the locks break and the actin filaments can slide out from the A bands again and the muscle becomes extended. Thus ATP is necessary both for contraction and relaxation of muscle. It is a very interesting fact that the enzyme ATPase which is responsible for the dephosphorylation of ATP is actually the myosin molecule itself.This is a very interesting example of a structural molecule which is also performing a metabolic function as an enzyme.

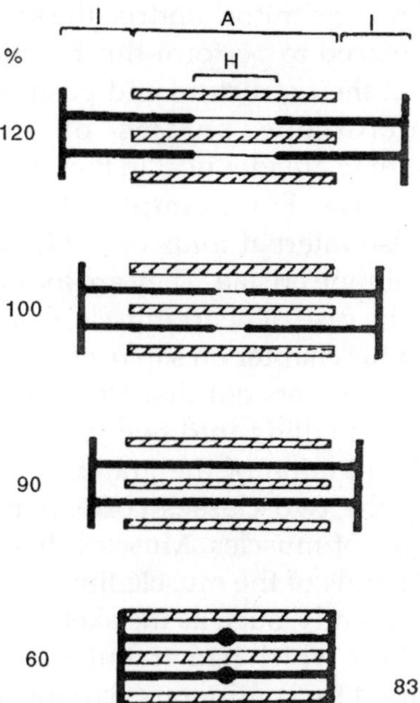

Fig. Molecular Basis of Contraction of Muscle

One should mention here just briefly one or two of the important chemical reactions which take place in muscle and which play a part in the contraction cycle. We have seen that the sliding of the actin filaments into the myosin filaments is dependent on the periodic formation and dephosphorylation of ATP. The re-formation of ATP is a result of the famous Lohnman reaction: creatine phosphate + ADP = creatine + ATP. The ATP, as has been mentioned before, is a high-energy phosphate and its dephosphorylation results in the liberation of energy which can be used for the mechanical movements necessary by the molecules in the contraction cycle. We do not propose to give full details here of the chemical reactions that take place in muscles in contraction.

We might note that the rephosphorylation of ATP requires energy to reconstitute a high-energy bond and this energy can be provided by oxidative phosphorylation produced as a result of coupling respiration with such a process. There are elements in the muscle which are capable of doing this and these elements are the muscle mitochondria, the sarcosomes, which are strategically placed to perform this function. We will say a word now about their structure and position.

Electron-microscope studies of the sarcosomes demonstrate that they are identical in their fine structure with that of mitochondria. For example, they have a double membrane and also internal folds of the inner membrane to form double membrane cristae. They are located in the muscle fibers, at least in the case of heart muscle, exactly opposite the A bands. Slater in his chapter on sarcosomes in "Structure and Function of Muscle" points out that Holmgren had originally classified muscles as falling into two types which depended on the type of distribution of the granules (sarcosomes) and he believed that the two classes corresponded to the two physiological types of muscles. Muscles that had granules at the level of the I bands of the muscle fibrils were those which acted only intermittently, such as the skeletal muscles of most vertebrates and those which had granules at the level of the A bands were required for continuous activity and in this group would come the flight muscles of birds and insects and, of course, the heart muscles of vertebrates.

It is of interest that, for instance, in the flight muscles of the humming bird and of the hovering insects where there is a phenomenal rate of contraction cycles per second, the mitochondria are relatively enormous in comparison with any other structural elements in the fibre, indicating the high rate of oxygen consumption that must be necessary when the fibers are contracting at such high speeds. Also interesting to note is that sarcosomes, like the sarcoplasmic reticulum, were discovered a good many years ago, in the 19th century in fact. Regaud in 1909 had claimed that they were, indeed, identical with mitochondria because they showed similar staining reactions. Yet, in histology books published since then and up to recent times, sarcosomes have been largely ignored in accounts of the histological structure of muscle. Thus, there are two curious facts about the history of muscle structure-- the sarcoplasmic reticulum and the sarcosomes were both discovered more than half a century ago and then dropped out completely from the histological picture and from the textbooks.

Not only do the sarcosomes resemble the mitochondria of other cells in staining reactions and in their fine structure, but they also contain the same series of respiratory enzymes, i.e., those enzymes concerned with the Krebs cycle and cytochrome system found in mitochondria. It is of interest that in red muscle fibers, where, as had been mentioned the red colour is due to myoglobin, there are more mitochondria than in white fibers and that the myoglobin assists in the rapid transfer of oxygen from the blood to the respiratory enzymes found in the mitochondria.The chemical reactions which take place in muscle can be stated as follows:

$$C_6H_{12}O_6 + 2\ DPN{+} + 2\ ADP + 2\ P\ (inorganic)$$
$$\rightarrow 2\ CH_3 \cdot CO \cdot COOH + 2\ DPNH + 2\ H + 2\ ATP\ (I)$$
$$2\ CH_3 \cdot CO \cdot COOH + 5\ O_2 + 30\ ADP + 30\ P\ (inorganic)$$
$$\rightarrow 6\ CO_2 + 4\ H_2O + 30\ ATP\ (II)$$
$$2\ DPNH + 2\ H{+} + O_2 + 6\ ADP + 6\ P\ (inorganic)$$
$$\rightarrow 2\ DPN{+} + 2\ H_2O + 6\ ATP\ (III)$$

Reaction I appears to be restricted to the sarcoplasm, in non-muscle cells it is in the cytoplasm; reactions II and III take place in the sarcosomes as they do in the mitochondria. If we sum up those three reactions, we see that the sum of the activities of these processes in the muscle is the synthesis of 38 molecules of ATP for each time the cycle occurs.

$$C_6H_{12}O_6 + 6\ O_2 + 38\ ADP + 38\ P\ (inorganic)$$
$$\rightarrow 6CO_2 + 6\ H_2O + 38\ ATP$$

The function of the sarcosomes in the muscle fibers appears to be the provision of ATP for the functioning of the contractile elements. The myofibrils have no membrane, the sarcosomes are virtually in contact with the myofibrils and the ATP can diffuse almost instantaneously into the sarcofibrillar filament. If an inadequate supply of oxygen is available or in muscles with an insufficient amount of sarcosomes, the DPNH and the pyruvic acid which are formed as a result of reaction I produce lactic acid. This reaction is

$$CH_3 \cdot CO \cdot COOH + DPNH + H{+}$$
$$\rightarrow CH_3CHOH \cdot COOH + DPN{+}$$

It is also of interest that sarcosomes can bring about oxidation of fatty acids and in the oxidation of stearic acid as much as 147 molecules of ATP are formed with each reaction. Also sarcosomes, similarly to mitochondria and other cell organelles, may become unstable if ATP is not present and, according to Slater, this suggests that some energy provided by the ATP is necessary to stabilize the structure of these organelles. It is believed that ATP is also used to some extent to maintain a difference in the concentration of ions as between the sarcosomes and the rest of the cytoplasm. Here then, in the muscle fibre we have probably a division of labour as well defined as one could find in any cell in the body.

There is a special part of the cell, the contractile part, which is exclusively used for this purpose, not only that but we have the contractile protein acting as an enzyme which can trigger off the whole process. The ATP which is necessary for the process is provided by the oxidative phosphorylations coupled with the respiration of the mitochondria which are themselves in close physical contact with the fibrils. The changes in ionic polarization of the membranes which are initiated by the motor endplate and which pass along the sarcolemma can be carried into the fibrils of the muscle fibre by means of the sarcoplasmic endoreticulum. Furthermore, the sarcolemma appears to be adapted to undergo pinocytosis and can take macromolecules directly into the sarcoplasm. This is the perfect example of division of labour in cells and yet it should be pointed out again how much these individual labors depend on and are integrated with each other.

THE NERVE FIBRE

In conclusion, we will consider the nerve cell and the nerve fibre. The nerve fibre itself is not a cell but it is a prolongation of a cell. It is a process of a neuron which is adapted for carrying nervous impulses. The functioning of the nerve fibre is a little more mysterious at the moment than is, for instance, the muscle fibre since we have little in the way of moving parts to investigate and a great deal of the information which is available is theoretical. Most of the studies on the physiology of the nerve fibre are devoted to the movements of ions, particularly sodium and potassium ions, in and out of the nerve fibre membrane since it is believed that it is the polarization and depolarization of the membrane produced by these ions which constitutes the passage of a nerve impulse along the fibre. The nerve fibre originates from the neuron, from a conical area known as the cone of origin. The nerve cell itself contains a welldefined nucleus and nucleolus. There are numerous Nissl bodies.

Among the characteristic structures classically described in the nerve cell are:

- Neurofibrils,
- Nissl bodies also called tigroid bodies because of their striped appearance which suggested a resemblance to the coat of a tiger to earlier investigators,
- Golgi apparatus,
- Mitochondria,
- Fats, phospholipids,
- Pigments.

Neurofibrils have been claimed to be present in all nerve cells and in the axon, but there is some doubt as to whether they exist as structural elements in the living cells although up to the beginning of the century they had been regarded as the basis of nerve impulse propagation. There is evidence from microdissection that in the living nerve cells of invertebrates they exist as discrete threadlike structures which are interwoven with each other and appear to be more viscous than the surrounding neuroplasm. In fixed and stained preparations, the neurofibrils are extremely fine and it is not surprising therefore they are hard to see in the living cell.

Fibrous structures have, however, been seen in the cytoplasm of nerve cells in tissue cultures and, if fresh nervous tissue is treated with salts such as $CaCl_2$ or various fixatives, the neurofibrils rapidly come into view. One view is that the neurofibril is a rodlet sol, that it is composed of orientated molecular particles which are not, in fact, sufficiently strongly attached to each other to form a definite fibril, but that a variety of agents may cause these rodlets to precipitate as fibrils.

Electron-microscope preparations also show Nissl bodies in nerve cells. They may be stained by basic aniline dyes and in specially fixed nervous material, histochemical tests have shown that they contain phosphoric acid and iron and they also appear to contain ribonucleic acid. In certain acute pathological conditions the Nissl bodies may disappear. This is a process known as chromatolysis and, if there is injury to the axon, the appropriate nerve cells show central chromatolysis, that is, the dissolution of the Nissl bodies surrounding the nucleus. There is also a loss of Nissl material from the central neurones of birds in long transmigratory flights. Under ultraviolet light Nissl bodies have the appearance of a flocculant precipitate and it has been shown that they can be displaced in the cell prior to fixation by ultracentrifugation.

Classic Golgi preparations show that the Golgi apparatus exists as a network close to and wholly or partly surrounding the nucleus, in fact, it was in nerve cells, as we have stated earlier, that Golgi first discovered the apparatus which is called after him.

Mitochondria are fairly obvious in properly stained nerve cells and in cell bodies they tend to be spherical but in the dendrites they are more elongated while in the axon they may appear as long filaments. Near the nucleus there may be an accumulation of fatty-like droplets which may be Golgi material and in a similar position two types of pigment may also be found, e.g., in the substantia migra of the midbrain a black-to-red melanin pigment occurs in this site in the cell. Many neurons show the presence of yellow droplets of lipochrome pigment.

The cell membrane of the nerve cells is an extremely delicate structure much finer than the nuclear membrane. Electron microscopy demonstrates that it cannot be more than a few hundredths of a micron thick, whereas the nuclear membrane is about 1/10 ? thick, probably due to its nature as a double fold of the endoplasmic reticulum. There is some evidence from the electron microscope that the cytoplasm of nerve cells is of high water content and that the material of the axon is still more dilute. The axon itself arises from an elevated portion

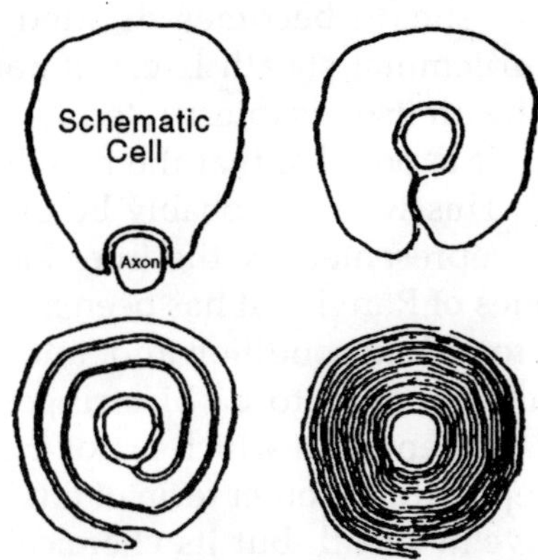

Fig. Schematic Representation of the Membrane Theory of the Origin of Nerve Myelin

of the neuron calles the axon hillock or cone of origin and it may give off branches or collateral but it usually ends in arborizations known as telodendria. A short way from the cell body, the axon develops a number of sheaths and becomes a nerve fibre.

If it has a well-defined myelin sheath, it is known as a myelinated nerve and if not, as an unmyelinated nerve fibre. Studies with polarized light however and with the electron microscope lead us to believe that all nerve fibers possess some type of lipoprotein sheath similar to the myelin sheath. Studies of the myelin sheath using polarized light, x-ray diffraction and later electron microscopy (particularly in the last case the work of FernandezMoran should be mentioned) have shown that lipoid and protein are laid down essentially in a series of alternating concentric layers. These lipoprotein layers appear to be about 130 A thick and the protein sections to be about 30 A thick. The lipid and the protein molecules of which these layers are composed are oriented radially, that is at right angles to the surface of the fibre.

When hematoxylin and eosin preparations are made of nerves fixed by conventional fixatives in the usual way and embedded in wax, a reticulate structure can be seen in the myelin sheaths which is known as the neurokeratin network. We shall have more to say about this later. Every now and again the myelin sheath becomes divided at the node of Ranvier, the neurilemmal sheath is continuous at this point but the myelin sheath itself is interrupted. The reason for this is not known but it is possible that the myelin may be a type of liquid crystal. This would probably be unstable in a tube longer than that represented by the internodes between the two adjacent nodes of Ranvier. It has been mentioned that the axon is also in a semifluid condition and, under the polarizing microscope, it can be seen to be birefringent suggesting a regular molecular orientation which is not incompatible with fluidity since it represents another form of liquid crystal similar to that of the myelin sheath but its chemical composition is different from that of the latter.

It has been found that the concentraton of ions is different inside and outside the cell and the nerve fibre; sodium and chloride, for instance, have a higher concentration extracellularly and potassium intracellularly. An electrical potential exists across the cell membrane because of this difference. Both sodium and potassium diffuse across cell membranes so that presumably they would eventually become equally distributed on either side of the membrane and the membrane potential would then disappear. But this does not happen and a factor which prevents it is a metabolic "pump" which maintains the difference in ion distribution. The level of the membrane potential depends upon the ratio of (K+) and (CI-) ions and the rate at which the sodium ion (Na+) is pumped.

The presence of a membrane potential along a nerve membrane is essential for the passage of an impulse. In the resting condition the membrane is known as the K+ membrane. During the passage of an impulse the potential across the nerve membrane changes, this is known as the action potential and is due to a change in the permeability of the membrane and a sudden influx of sodium ions. The nerve fibre membrane is then called a (Na+) membrane. The recovery of the membrane and the reestablishment of a (K+) membrane starts with an increase in permeability to (K+) ions and a decrease in permeability to (Na+) ions. These changes finally bring the potential back to the resting value.

The studies on the conductance of (Na+) and (K+) ions in nerve were carried out by Hodgkin and Katz with the introduction of the membrane voltage clamping technique. In effect what they demonstrated was that the movement of (Na+) into the fibre during the passage of an impulse depolarizes the membrane by carrying electrical charges through it. The reestablishment of a resting potential across the fibers (the K membrane) requires the sodium ions to be pumped out against a gradient of (Na+). This can only be done by the use of energy. That metabolic processes are involved in this action is indicated by the relatively high oxygen uptake of nerve.

By using optical methods, it has also been demonstrated that changes in the electrical activity of the nerve fibre which are associated with the passage of a stimulus are associated with changes in the optical density or in the light scattering capacity of the fibre. The scattering of light by the fibre indicates change of volume. It is of interest that this swelling is maintained for several seconds after the summated action of several series of impulses. The rate of change in optical density is not comparable to the velocity of propagation of the action potential, but there appears to be a connection between the wave of excitation and the optical changes. However, the nature of this connection is obscure at the moment.

The fluid substance of the axon is believed to be under pressure. As evidence for this one might quote the experiments of Paul Weiss in Chicago, who found that if a nerve fibre is ligated, the axoplasm is dammed up to the point where the constriction occurs and when pressure is released it flows forward again. He also showed that, if a nerve fibre is cut, the axonal material will stream out from the cut surface. It has also been shown that, if such a severed nerve is stimulated, the speed of outflow of axoplasm is decreased. This appears to be due to gelatinization of the axoplasm and a consequent increase in its viscosity.

In our laboratory at Emory University Professor Portela

and his colleagues have been making studies of the correlation of physicochemical properties of the axoplasm in a state of excitation. Studies are being made with an interference microscope of surface movements of the nerve in response to "stimulation" and these movements are being recorded simultaneously with change in action potential. The surface movements of the nerve are indicated by the reflection of interference fringes in the interference microscope of the light from the nerve surface.

It is interesting to note that the Russian authors Kayushin and Lyudkovskaya believed that the surface movements (as expressed by shifts in the interference bands) result from changes in the fibre volume and since these shifts disappeared when there was no stimulation, it appeared that these volume changes had returned to normal and the movement was probably of an elastic nature. These mechanical waves (surface movements, volume changes) are presumably connected with changes of the molecular micellar structure in the nerve. Thus, it appears that the whole of the structural system of the nerve takes part in the conduction of an impulse. This way of looking at the relationship between molecular structure and the excitationconduction-recovery cycle in the nerve is a new approach to the investigation of nerve physiology.

It is possible that the mechanical waves in the nerve fibre push acetylcholine down toward the myoneural junction and there is some evidence this compound may be formed either in the axon or possibly in the cell body itself. There is evidence of considerable synthesis of protein by nerve cells but the significance of this is not known.

In our laboratory attempts are being made to correlate the lipid and protein structure and organization of the nerve and function of the nerve as a conductor. The approach is similar to that of Tobias. Proteolytic and lipolytic enzymes are used to produce structural changes in the nerve fibre and afterward records of interference and electrical measurements and fine structural changes are made. The removal of phospholipids from the nerve membrane results in the nerve becoming inexcitable but, following treatment with proteolytic enzymes, nerve conduction is not blocked. It is of interest that pressure (about 11½ lbs.) on the nerve will cause an interruption of the normal function of a motor but not of a sensory nerve and this may last some two weeks before coming completely back to normal. This is particularly curious when such pressure produces no microscopically visible degenerative changes in the nerve fibre.

The cell membrane in nerve fibers is complicated because of the close association of the axon with the Schwann cell. The Schwann cell is associated with the axon in both myelinated and nonmyelinated fibers. In the case of the nonmyelinated fibers, the axon is embedded to varying degrees into the protoplasm of the Schwann cell, but is still surrounded closely by the cell membrane of the Schwann cell which it has invaginated with it. The actual distance between the outer parts of the two membranes under these circumstances is approximately 150 A. The membrane of the axon and the membrane of the Schwann cell are each approximately 75 A and are made up of three parts. There are two rather dense areas which are about 25 A across and these are separated from each other by a less dense area also about 25 A across. So that we may consider that the membrane of the nerve fibre is enlarged to this extent by the membrane of the Schwann cell itself.

In myelinated fibers the axon instead of just being embedded in the cytoplasm of the Schwann cell is surrounded by many layers of Schwann cell cytoplasm and cell membrane. The mechanism which produces this has been demonstrated by Geren in tissue cultures of developing nerve fibers. She demonstrated that the Schwann cell winds in a spiral fashion around the axon thus building up many layers of Schwann cell membrane around the axon itself. Many degrees of myelination of nerve fibers are known and the degree of myelination depends on the number of times the Schwann cell winds around the axon.

Electron-microscope studies of myelinated nerve fibers have demonstrated very complex arrangements of the myelin so produced. The first of the high-resolution studies of such fibers were produced by Fernandez-Moran and Sjöstrand. If a myelinated nerve is first fixed in osmium tetroxide (osmic acid), it could be seen that the myelin was made up of a series of lines which absorbed the osmium very strongly. These lines were approximately 25 A thick and repeated at a period of up to 120 A. Each of these dark lines was separated by a light line of approximately the same size and running across the centre of each light line was a line which was more dense than the light line but less dense than the main 25-A repeating lines. In interpreting the significance of these lines, Robertson has pointed out that the gap, which in the unmyelinated fibers was between the two cell membranes and measured 75 A across, is largely eliminated in myelinated fibers and the outer part of the membrane of the Schwann cell comes into direct contact with the membrane of the axon.

Thus we can say that, as Robertson points out, each of the repeating units which are obvious in electron micrographs of the myelin sheath is two Schwann cell membranes in contact along their outside surfaces. The Schwann cell membrane probably consists of a single bimolecular leaflet of lipids of which the hydrophilic ends are associated with monolayers of a nonlipid material and which is probably protein, although there is a possibility that either some polysaccharide or glycoprotein of some sort may be associated with it, possibly on the side directed toward the cytoplasm.

It has recently been shown by Tewari and present author that, in addition to the complex structure demonstrated by the electron microscope, the neurokeratin network previously considered to be a fixation artifact of myelin may indicate the presence of some differentiation of the myelin sheath in the living nerve fibre and the presence of various enzymes including a large variety of phosphatases as well as ATPase and creatine phosphatase (acid phosphatase is restricted to the axon). There is also a good deal of phospholipid as demonstrated by the Baker phospholipid technique. This network has been found, in fact, from these preparations to consist of areas resistant to damage by fixatives and other reagents and apparently arranged in the form of hexagonal prisms which radiate from the axon to the surface of the fibre (actually to the neurilemma) and that it is along these areas that the enzymes and the phospholipid we have mentioned appear to be concentrated. The areas appear to contain at least unsaturated fat and lipid and it is of interest that, in nerve which has been osmicated for 5 days and extracted with turpentine, it is the material from the interior of these active areas that is extracted, whereas the similarly osmicated material on the faces of the active areas remains.

It is possible that the reason why it is retained in the faces of the prisms is due to the fact that the osmicated lipids are bound strongly to a protein and cannot be extracted by turpentine. There is also some evidence that it is held in place by polysaccharide. It is of interest, too, that in these same areas techniques which would normally demonstrate mitochondria and which are presumably really demonstrating a variety of lipoprotein show a localization of similar compounds in the neurokeratin network. It is difficult to reconcile this work with that of electron microscopy. Although if we assume that the protein of the myelin lamellae is enzyme protein in the regions of the neurokeratin network but not in between them, then we can reconcile the two suggested types of structure. We also have to ask what happens to the mitochondria of the Schwann cell when they wrap around the axon.

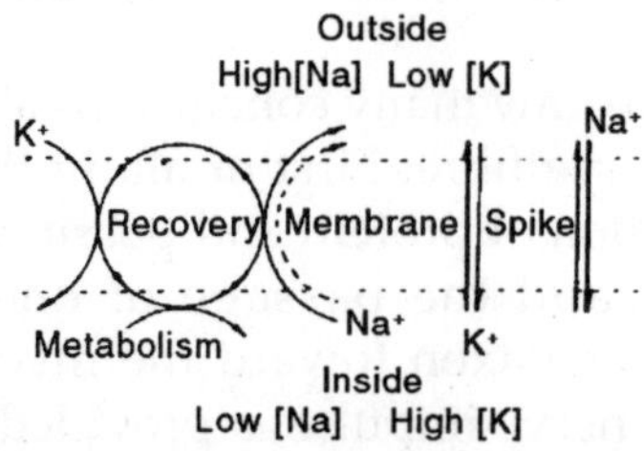

Fig. Diagrammatic Illustration of Ion Movements Through the Nerve

Do they disintegrate or do they become enormously elongated so that their membranes constitute some of the lamellae described by the electron microscopists.

We have mentioned that there is a high concentration of potassium ions inside the nerve fibre in the axon and much less concentration of sodium. Now it appears that sodium continuously diffuses into the nerve fibre and it must be excreted from the axon to the exterior again. The difficulty about this is that the sodium has to be excreted against a gradient, that is to say, sodium has to be passed out of the axon into a medium where there are more sodium ions than there were in the axon and vice versa with potassium. This process requires energy and there have been two principal views as to how this could be done. The two theories are the redox pump and the sodium pump theory.

Conway has suggested the following way in which the outward movement of sodium ions in the nerve fibre could be explained. Sodium ions could be continuously passed on from unsaturated phospholipid molecules to the next unsaturated phospholipid molecules through the membrane to the outside and there would be alternate oxidation and reduction of the phospholipid molecules as this happened.

The other view is that sodium is pumped out by energy produced by presence of ATP, creatine phosphate and oxidative enzymes and possibly also with the aid of phospholipid. What is not certain yet is whether Na and K ions need to be passed through the myelin sheath to maintain the physiological functioning of the nerve fibre. This is a problem for which we will no doubt get the answer in due course.

It is difficult to draw many conclusions about the division of labour in the nerve fibre. First of all, in the nerve cell we have great production of protein and possibly the production of acetyl choline and the passage of one or both these compounds down the axon toward the myoneural junction; the passage of the nerve impulse is provided for by the axon membrane itself.

In myelinated nerves we may have the metabolic mechanism of the neurokeratin network which might be used for pumping ions away from the area of the axon membrane. However, the precise relationship of the myelin sheath to the propagation of the nerve impulse is not known. In myelinated nerves, the impulse appears to be faster than in nonmyelinated or poorly myelinated nerves, anti one does not know quite in what way the myelin sheath is able to do this.

At the termination of the nerve fibre there is a modification in the form of a complex branching in association with the sarcolemma to build up the myoneural junction. There is an accumulation here of acetyl choline which presumably passes through the membrane of the neural part of the mechanism to stimulate the formation of a contraction wave by its influence on the sarcolemmal membrane. The presence of cholinesterase in these junctions, which is necessary for inactivation of acetyl choline after it has done its job, is of interest but its precise origin is unknown. So we can see evidence of division of labour in the nerve cell and its axon but the nature of the labour performed by each part is still partly speculative.

We have studied the structure and to some extent the chemical composition and the biochemical composition of the cell in general and of three specialized types of cells. We have demonstrated that various parts of the cell perform very important but different functions, that, in other words, there is a division of labour. Each part performs the job for which it is chemically and structurally suited; thus the nucleolus synthesizes a considerable amount of protein and ribonucleic acid. We know that the chromosomes of the nucleus contain the genes which produce hereditary effects in the cell. We have seen that the oxidative activities of the cell are carried on in the mitochondria and that, as a result of respiratory activity of these structures, ATP is produced which is available for energy and structural purposes in the cell. In the glandular cell, the ATP provides energy for the synthesis of the products of secretion.

In the muscle fibre, it provides the energy for muscular contraction; and presumably in the nerve fibre the myelin sheath provides the energy for the extrusion of Na ions to the neurilemma (although in this latter case this is an example of division of labour between cells. The cytoplasmic (endoplasmic) reticulum is still a very debatable structure; the presence of ribonucleoprotein granules attached to the outside of these membranes indicates that they are probably regions of protein synthesis. If, as is believed by some authors, the cistemae in the endoplasmic reticulum are continuous with the outside we have, in fact, an enormous increase of area of the membrane of the cell. We have the outside world actually penetrating deep into the cytoplasm and even surrounding the nucleus so that nuclear products could be secreted directly to the exterior of the cell without actually passing through the cytoplasm itself. In this way glucose could pass directly into the nucleus as well as directly into the interior of the cytoplasm.

What is so striking in the cytoplasm is not that there is a division of labour but, what is more fantastic, the degree to which all the various parts of the cell cooperate with each other to build up a metabolic picture which is ordered and controlled. This co-operation rarely gets out of gear and when it does disorders of growth and metabolism occur of which one variety can be cancer. The mechanism of cell function is under control of the endocrine system and we have mentioned earlier in this book how the fine structure of the prostate cell is under control of the male sex hormone and how the localization of acid phosphatase in the same cells is also under control of the same hormone. This is a field which is still scarcely touched and is likely to produce in the future an enormous amount of interesting and valuable information about the function of the cell.

The relationship of vitamins and nutritive factors in general to cell structure is also only beginning to be investigated. Studies on starvation have demonstrated a great increase in the number of mitochondria and our own studies on scurvy have demonstrated the more intimate association of the endoplasmic reticulum with the mitochondria with a simultaneous increase in the number of mitochondria. Here is a vast field in nutrition where electron microscopy would probably produce immense contributions to a study not only of the physiology of the cell but of the function of vitamins and other nutrients in cell metabolism.

The precarious nature of the cell structure is demonstrated by the fact that the structural integrity of mitochondria is affected in the absence of ATP and it seems possible that this whole complicated system of energy production is essential not only to provide for cell activity but also energy to maintain the structural integrity of the cell and its organelles; if this is so, then any serious interference with the ATP production cycle will lead to a breakdown of the complex cellular structure as well as interference with the metabolism.

New techniques of cytological investigation have in the last few years made fundamental alterations in our outlook on cell structure and function and we can expect that the next few years will shed much more light on the intriguing subject of the division of labour in cells.

Chapter 8

The Endoplasmic Reticulum

In 1945, just as the electron microscope was becoming really useful as a research tool, Albert Claude of Belgium and Keith Porter at the Rockefeller Institute discovered a vast network of channels bounded by membranes in the cytoplasm of chick embryo cells. At times these looked like the layers of an onion. Porter, who is now at the University of Colorado, called this network the endoplasmic reticulum because it was more concentrated in the inner (encloplasmic) region of the cell than in the peripheral (ectoplasmic) region. Similar networks were later found in nearly all eucaryotic cells.

It turned out that the membranes of this endoplasmic reticulum (ER) all interconnect, forming a system of tubes and flattened sacs which is continuous with the nuclear membrane. In effect, this system divides the cytoplasm into two main regions: one enclosed within the "plumbing," and the other forming the outer region, or cytoplasmic matrix.

Some of these membranes are smooth (the SER). Others are "rough" (the RER), dotted with ribosomes which form granules on their outer surfaces. The outer layer of the nuclear envelope, with which they connect, also contains bound ribosomes which may synthesize special classes of proteins.

The ER's chief function appears to be the storage, segregation and finally transport of substances (mostly proteins) which the cell manufactures for use in other sections of the cell or outside the cell. These proteins are synthesized on ribosomes that are bound to the RER.

In the middle 1950's, George Palade concluded that the size of the RER in a cell corresponds pretty closely to the quantity of protein which the cell exports. Plasma cells which produce antibodies have highly developed RERs with large storage cavities, for example. So do fibroblasts (connective-tissue cells) which produce collagen and pancreatic cells which produce digestive enzymes. The gamma globulin produced by plasma cells is found mainly in these storage cavities, from which it goes forth to combat specific infections.

Cells that manufacture proteins just for their own use have little or no RER, however. In such cases, the protein is synthesized on free ribosomes and circulates freely in the cytoplasm. While the attached ribosomes of the RER are primarily concerned with manufacturing proteins for secretion, the membrane of the whole ER synthesizes lipids (fats), including cholesterol. It is particularly well developed in certain cells where it takes on some extra function -- for example, in liver cells, where it breaks down (metabolizes) drugs.

Researchers have had a hard time understanding exactly how the ER works, partly because it is impossible to isolate it intact with an ultracentrifuge -- the ER simply breaks up into fragments (these are called microsomes). However, they are beginning to make some progress. Recently, for instance, Gunther Blobel of Rockefeller University discovered a "signal sequence" on messenger-RNA which determines whether the ribosomes it interacts with will attach themselves to the ER's membranes and make secretory proteins. The same sequence then leads the newly synthesized protein to storage cavities in the RER.

THE GOLGI APPARATUS, A PACKAGER OF PROTEIN

For a long time biologists disagreed about what it was that Camillo Golgi, an Italian scientist, had actually seen under his light microscope in 1898. Golgi had taken nerve cells from a barn owl and a cat, stained them, studied them and noticed what looked like a separate structure, distinct from the nucleus, in the cytoplasm. This structure clearly took on a different colour from the rest of the cell. However, some biologists thought it was just an artifact -- perhaps something related to the chemicals he had used.

Half a century later, electron miscroscopists confirmed that this structure -- now called the Golgi apparatus -really exists and recently it has been shown to play a most important role in the packaging of proteins for export from the cell. The Golgi apparatus consists of stacks of flat, membranous sacs that are piled one on top of the other. It receives molecules of protein from the ER's cavities; wraps up large numbers of these molecules into a single, membranous envelope; and sends the package on its way to the cell's surface.

The wrapping around these packages allows the cell to keep large concentrations of proteins on the ready. Some enzymes such as trypsin, which helps to break down foodstuffs in the intestinal tract, could have a disastrous effect on other proteins in the cell that made them unless they were carefully segregated or stored in an inactive form. In the 1960's, researchers showed that such enzymes (in their inactive form) are enclosed in vesicles (capsules) in the Golgi region.

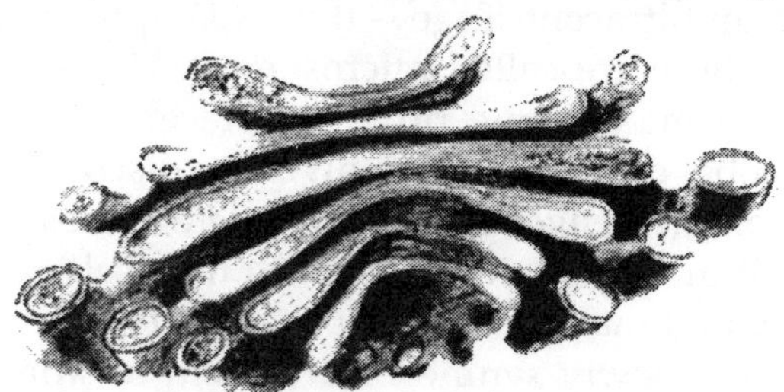

Fig. The Golgi Apparatus

The vesicles then migrate to the cell's surface, where they fuse with the cell's outer membrane and release their contents. This is how cells secrete their hormones, enzymes or other types of proteins when needed. Before packaging these proteins, however, the Golgi apparatus sometimes processes them further. It may add carbohydrates to their molecules, as it does in certain mucus-secreting cells whose Golgi apparatus is so big that it dominates the entire cell. Or it may remove part of a polypeptide chain, as in the case of insulin, which is processed from a low-activity precursor to a fully active hormone in the Golgi complex. In addition to preparing proteins for export, the Golgi apparatus appears to be responsible for concentrating and wrapping certain enzymes into separate organelles -lysosomes -- which remain inside the cell.

LYSOSOMES, THE CELL'S SCAVENGERS

When a white cell engulfs a bacterium and destroys it, the white cell's lysosomes do most of the work: They release their digestive enzymes around the engulfed material and that is what breaks it down.

Similarly, when a cell takes in large molecules of foodstuffs, the lysosomes break these molecules down into smaller and simpler products which can be used by the cell. These products diffuse through the lysosome's membranes and go into the rest of the cell, where they serve as building blocks for various structures, until nothing is left inside the lysosome but undigestible material -- and the lysosome becomes what is called a residual body. In some cells the residual body then migrates to the cell surface and gets rid of the undigested material by ejecting it into the external environment; in crude terms, it acts as the cell's garbage disposal system.

Lysosomes were discovered by a Belgian researcher, Christian de Duve, in 1949, when he homogenized (ground up) some animal cells and separated them into various fractions by running them through an ultracentrifuge. After one of these fractions had been left standing for a few days, he noticed that the level of a certain enzyme in it rose dramatically. Since this enzyme had not attacked any part of the cells before they were ground up, he reasoned that it must have been kept segregated from the rest of the cell -- probably inside some kind of organelle. He also knew that he had used a relatively gentle method of homogenization, which could have allowed the unknown organelle to remain intact; presumably it released its contents later.

De Duve's biochemical approach, for which he shared the Nobel Prize with Claude and Palade in 1974, was soon supplemented by electron microscopy. But it proved extremely difficult to identify the new particles, since they had no characteristic shape or internal structure. They seemed to change size and shape constantly, according to their activity. Finally, in 1955, Alex Novikoff of Yeshiva University clearly identified some lysosomes in rat liver cells and it is now believed that lysosomes exist in all animal cells. Something very similar to lysosomes also exists in plant cells.

More than 30 hereditary diseases have now been linked to the absence of various digestive enzymes in some persons' lysosomes. In each case the missing enzyme leads to the pile-up of a different kind of waste material. In Tay-Sachs disease, for instance (a disease which occurs mainly among Jews of eastern European descent), an enzyme deficiency caused by a single gene results in progressive damage to brain and other cells and to death by the age of 4.

There is still no treatment for such "storage diseases " -- diseases accompanied by an accumulation of waste materials in the victims' cells. About a decade ago, however, Elizabeth Neufeld of the NIH showed that these defects could be observed, studied and even corrected in a test-tube culture of cells taken from patients with these diseases. The corrective factors which she supplied were the specific enzymes which these patients' cells lacked.

This led to the idea of treating patients in a similar way, with enzyme replacement therapy. However, it was not known how the missing enzymes could be delivered to the patient's cells in all parts of his body, including his brain, nor what would induce the lysosomes in each cell to take them up. When purified enzymes are injected directly into body fluids, they tend to be quickly destroyed or inactivated and they are not taken up by the tissues which need them most. It is particularly difficult to make such enzymes cross the blood-brain barrier -- an important problem, now being investigated, since several lysosomal diseases produce severe mental retardation.

Recently Gerald Weissman of New York University developed a promising technique which may fool the white blood cells and other scavenger cells (possibly including the brain's glial cells) into accepting the missing enzyme. Mixing lipids (fatty substances) with a solution of the missing enzyme, he shook up the mixture and manufactured microscopic bubbles that contained the enzyme. Next he coated these bubbles, or liposomes, with antibodies to make them attractive to white blood cells. When he fed these coated liposomes to Tay-Sachs cells in a test tube, he was happy to find that the white blood cells devoured the liposomes and absorbed their enzymes. Many questions remain about the potential medical use of this technique, but it may lead to an effective treatment of lysosomal storage diseases.

A different line of research has focused not on the lysosomes' deficiencies, but on the way in which these organelles guard their contents. Lysosomes are sometimes called the cell's "suicide bags" because they contain enzymes that can digest almost anything in the cell: proteins, RNA, DNA and carbohydrates. Dangerous and corrosive as these enzymes may be, they do not damage the cell itself as long as the lysosome's membrane remains intact.

When cells are programmed to die in some normal processes of embryonic development, however -- for example, in the metamorphosis of insects -- the lysosomes' membranes become permeable and release their enzymes to digest the cells from within. In very old cells, too, the lysosomes may release their contents, which destroy the cell.

The same kind of "autodigestion" can occur to cells that have been injured by lack of oxygen, an excess of vitamin A, exposure to certain carcinogens, or starvation. Then the lysosomes' membranes suddenly become permeable or rupture and their enzymes leak out, to digest parts of the cell. Surprisingly, this autodigestion may be temporarily useful if the cell cannot get enough food from outside sources. The lysosomes then break down some of the cell's own contents, liberating some building blocks that can be used to make more essential substances and ensuring the cell's survival without major damage. At other times, however, the rupture of lysosomal membrances may cause the cell's death.

The inflammation and pain of arthritis may also be related to a leakage of lysosomal enzymes from certain white blood cells. (This link is supported by the fact that substances which reduce inflammation, such as cortisone, are known to strengthen the lysosomal membranes). In many ways, then, health and disease depend on the lysosomal membrane's ability to control the release of its contents. Some substances seem to stabilize and strengthen this membranous envelope, while others weaken it. In the future, researchers may find ways to use these properties for the prevention of disease or for therapy.

This is still a long way off, however. At present, much remains to be learned about how and where lysosomes are made (some lysosomes appear to come from the Golgi apparatus, others from the ER), what stages they go through and what controls their activities.

Chapter 9

The Dynamic Cell

To study the properties of the molecules of life and the innumerable variations on basic themes that are found in different organisms, modern researchers employ concepts and experimental techniques drawn from biochemistry, molecular biology, genetics, and cell biology. The resulting discipline of *molecular cell biology* investigates how cells develop, operate, communicate, and control their activities, and on occasion go awry. This book contains the authors' attempt to describe systematically the current state of knowledge about cells and to present many of the key experiments that have led to our current understanding of cellular life. It may seem to you, our new reader, a daunting challenge. Our hope is that the overwhelming inventiveness and sheer beauty of construction of biological systems will intrigue you and amply reward your efforts to understand the story we tell.

Living systems, including the human body, consist of such closely interrelated elements that no single element can be fully appreciated in isolation from the others. Organisms contain organs; organs are composed of tissues; tissues consist of cells; and cells are formed from molecules. The unity of living systems is coordinated by many levels of interrelationship: molecules carry messages from organ to organ and cell to cell; tissues are delineated and integrated with other tissues by noncellular membranes secreted by cells; and cells gain identity from contact with other cells. Generally all the levels into which we fragment biological systems interconnect. To learn about biological systems, however, we must take a segment at a time. The biology of cells is a logical starting point because an organism can be viewed as consisting of interacting cells, which are the closest thing to an autonomous biological unit that exists. The integration of cellular activity into tissues, the development of organisms by growth and specialization of cells, and the metabolic events fueling the dynamism of living systems are all topics on which we will touch, but they are all topics that fall within the province of other subdisciplines of biological science.

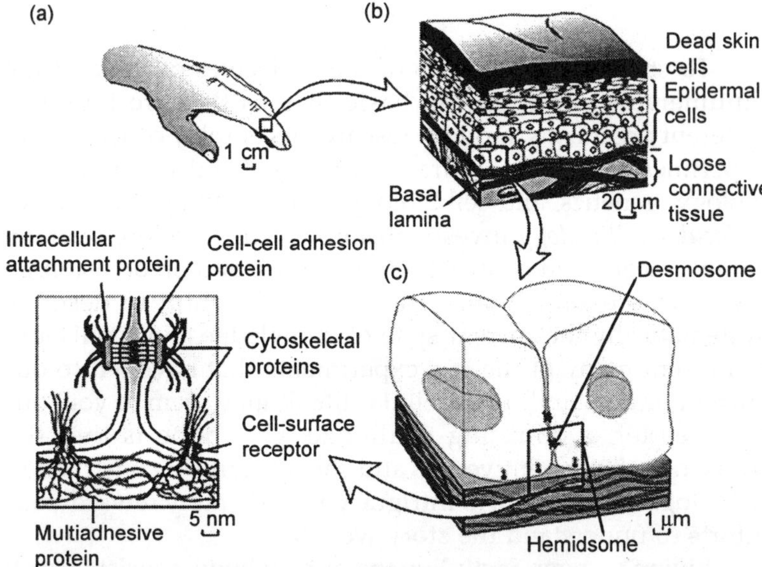

Fig. Living systems such as the human body consist of closely interrelated elements.

(a) The surface of our hand is covered by a living organ, skin, that is covered by several layers of tissue.

(b) An outer covering of hard, dead cells protects the body from injury, infection, and dehydration. This layer is constantly renewed by living epidermal cells, which also give rise to hair and fur. Deeper layers of muscle and connective tissue give skin its tone and firmness.

(c) Tissues are formed through subcellular adhesion structures that join cells to each other and to an underlying layer of supporting fibers.

(d) At the heart of the adhesion are its structural components: phospholipid molecules that make up the plasma membrane and protein molecules which form strong bonds with molecules on other cells and with internal or external fibers.

In this chapter, we provide a framework for understanding the primacy of cells in biological systems and review several fundamental concepts that recur throughout our more detailed discussions in subsequent chapters. We begin with a brief look at the role of evolution and then discuss the general properties of the molecules found in biological systems. Next, we review the main features of cellular architecture, noting the similarities and differences between the three main cell lineages that have emerged over evolutionary time. The remaining topics covered in this chapter focus on the assemblage of cells into organized structures and their dynamic nature, in preparation for later chapters dealing with various processes critical to cellular growth, differentiation, and adaptation to changing circumstances.

EVOLUTION: AT THE CORE OF MOLECULAR CHANGE

The interplay of events played out over billions of years, in the historical process called *evolution,* dictates the form and structure of the living world today. Thus biology, which is the study of the results of these historical events, differs fundamentally from physics and chemistry, which deal with the essential and unchanging properties of matter. The great insight of Charles Darwin was that all organisms are related in a great chain of being extending from the distant past to the present. The Darwinian principle that organisms vary randomly and the fittest are then selected by the forces of their environment guides biological thinking to this day.

We now know that genes, which chemically are composed of deoxyribonucleic acid (DNA), ultimately define biological structure and maintain the integration of cellular function. The genes encode proteins, the primary molecules that make up cell structures and carry out cellular activities. Alterations in the structure and organization of genes thus provide the random variation that nurtures evolutionary change in biological structure and function.

Even scientists brought up in the evolutionary tradition have been surprised to learn in recent years just how closely the genes of different species are related. During evolution, genes have been conserved to such an extent that some human genes will function in a yeast cell and quite a few will function in a fly cell. Clearly, one feature of evolution is the *maintenance* unchanged of many aspects of cellular life even while great changes in external form and capability are occurring. Recent progress in determining the sequences of all the genes in a variety of organisms is revealing the subtle changes that have fueled evolution.

The creative part of the evolutionary process is *adaptation* to rapidly changing environments and the conquest of new environmental niches. During this process, small alterations in cellular structures and functions are selected. Entirely new structures rarely are created; more often, old structures are adapted to new circumstances. More rapid change is possible by rearranging or multiplying previously evolved components rather than by waiting for a wholly new approach to emerge. The cellular organization of organisms plays a fundamental role in this process because it allows change to come about by small alterations in previously evolved cells, giving them new capabilities

THE MOLECULES OF LIFE

Among the many events that occur in the life of a cell are a multitude of specific chemical transformations, which provide the cell with usable energy and the molecules needed to form its structure and coordinate its activities. Here we briefly describe the functions of the main types of chemicals that compose cells. Throughout many later chapters we will focus on the interactions and transformations of these molecules.

Water, inorganic ions, and a large array of relatively small organic molecules (e.g., sugars, vitamins, fatty acids) account for 75 – 80 percent of living matter by weight. Of these small molecules, water is by far the most abundant. The remainder of living matter consists of macromolecules, including proteins, polysaccharides, and DNA. Cells acquire and use these two size classes of molecules in fundamentally different ways. Ions, water, and many small organic molecules are imported into the cell. Cells also make and alter many small organic molecules by a series of different chemical reactions. In contrast, cells can obtain macromolecules only by making them. Their synthesis entails linking together a specific set of small molecules (monomers) to form polymers through repetition of a single type of chemical-linkage reaction.

Some small molecules function as precursors for synthesis of macromolecules, and the cell is careful to provide the appropriate mix of small molecules needed. Small molecules also store and distribute the energy for all cellular processes; they are broken down to extract this chemical energy, as when sugar is degraded to carbon dioxide and water with the release of the energy bound up in the molecule. Other small molecules (e.g., hormones and growth factors) act as signals that direct the activities of cells and nerve cells communicate with one another by releasing and sensing certain small signaling molecules. The powerful effect on our body of a frightening event comes from the instantaneous flooding of the body with a small-molecule hormone that mobilizes the "fight or flight" response.

Macromolecules, though, are the most interesting and characteristic molecules of living systems; in a true sense the evolution of life as we know it is the evolution of macromolecular structures. Proteins, the workhorses of the cell, are the most abundant and functionally versatile of the cellular macromolecules. To appreciate the abundance of protein within a cell, we can estimate the number of protein molecules in a typical eukaryotic cell, such as a hepatocyte in the liver. This cell, roughly a cube 15 μm (0.0015 cm) on a side, has a volume of 3.4×10^{-9} cm^3 (or milliliters). Assuming a cell density of 1.03 g/ml, the cell would weigh 3.5×10^{-9} g. Since protein accounts for approximately 20 percent of a cell's weight, the total weight of cellular protein is 7×10^{-10} g.

The average yeast protein has a molecular weight of 52,700 (g/mol). Assuming this value is typical of eukaryotic proteins, we can calculate the total number of protein molecules per liver cell as about 7.9×10^{9} from the total protein weight and the number of molecules per mole, which is a constant (Avogadro's number). To carry this calculation one step further, consider that a liver cell contains about 10,000 different proteins; thus, a cell contains close to a million molecules of each protein on average. In actuality, however, the abundance of different proteins varies widely, from the quite rare cell-surface protein that binds the hormone insulin (20,000 molecules) to the abundant structural protein actin (5×10^{8} molecules).

Many of the proteins within cells are enzymes, which accelerate (catalyze) reactions involving small molecules. Other proteins allow cells to move and do work, maintain internal cell rigidity, and transport molecules across membranes. Proteins even direct their own synthesis and that of other macromolecules. Reflecting their numerous functions, proteins come in many shapes and sizes. The elucidation of the structure of proteins and the relation of protein structure to function remain active areas of scientific investigation. Proteins are formed from only 20 different monomers, the amino acids. That such a limited set of building blocks can do so much is a continuous marvel, even to researchers who work with proteins every day. They are the true glory of the biological world.

The macromolecule that garners the most public attention is not protein but deoxyribonucleic acid (DNA), whose functional properties make it the cell's master molecule. The three-dimensional structure of DNA, first proposed by James D. Watson and Francis H. C. Crick about 50 years ago, consists of two long helical strands that are coiled around a common axis forming a double helix. The double-helical structure of DNA, one of nature's most magnificent constructions, is critical to the phenomenon of heredity, the transfer of genetically determined characteristics from one generation to the next.

Each strand of DNA is composed of just four different types of monomers called nucleotides. Genes are simply coded representations of the structures of individual proteins, a code written in four chemical "letters" — the nucleotides — and displayed as a continually varying sequence in DNA. Since cells use proteins (enzymes) to make other molecules like sugars or fats, DNA indirectly directs the synthesis of many small molecules as well as proteins. DNA also contains a coded set of instructions about when various proteins are to be made and in what quantities.

In the common view, DNA is the storage form of genetic information, which protein "machines" read out for use by the cell. But a third macromolecule, ribonucleic acid (RNA), is necessary in the process. The *central dogma* of biology states that the coded genetic information hard-wired into DNA is transcribed into individual transportable cassettes, composed of messenger RNA (mRNA); each mRNA cassette contains the programme for synthesis of a particular protein (or small number of proteins). This critical trio of macromolecules — DNA, RNA, and proteins — is present in all cells. The mechanism whereby the information encoded in DNA is deciphered into proteins is now understood quite well and explained. How this process of gene expression is regulated — that is, how cells "know" to make the right proteins at the right time in the right amounts — is a major focus of current research in molecular cell biology and a recurring theme throughout this book.

THE ARCHITECTURE OF CELLS

Although generalizations in biology usually lack the theoretical underpinnings found in physics, there are very clear commonalities among living systems that give biology a unity. One is the style of cellular construction. The biological universe consists of two types of cells — *prokaryotic cells,* which lack a defined nucleus and have a simplified internal organization, and *eukaryotic cells,* which have a more complicated internal structure including a defined, membrane-limited nucleus. Detailed analysis of the DNA from a variety of prokaryotic organisms in recent years has revealed two distinct types: bacteria (often called "true" bacteria or eubacteria) and archaea (also called *archaebacteria* or *archaeans*). As we discuss the archaea are in some respects more similar to eukaryotic organisms than to the true bacteria.

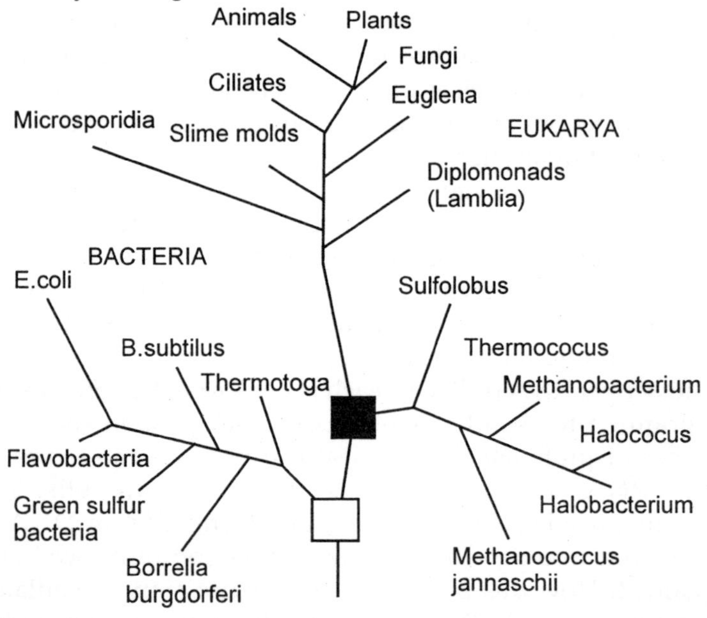

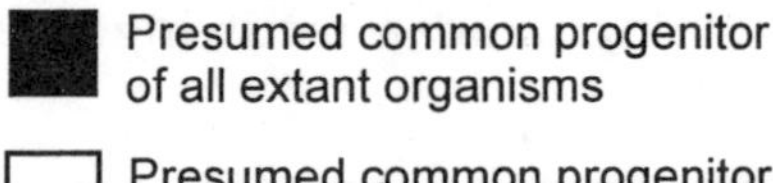

Fig. The three kingdoms of organisms are related through common sequences of their ribosomal RNAs.

Based on the assumption that organisms with more similar genes evolved from a common progenitor more recently than those with more dissimilar genes, researchers have developed the lineage tree. According to this tree, the archaea and eukarya (eukaryotes) are thought to have diverged from the bacteria before they diverged from each other. Despite the differences in the organization of prokaryotic and eukaryotic cells, all cells share certain structural features and carry out many complicated processes in basically the same way.

CELLS ARE SURROUNDED BY WATER-IMPERMEABLE MEMBRANES

A cell, because it is a limited space, must have an outer border. The construction of that border represents one of the most fundamental considerations in biological organization. The outer shell of cells, like any shell, is built to keep the interior contents from leaking out into the surrounding environment. The chemical processes of cellular life generally take place in a watery solution, and the intracellular constituents of cells are largely molecules that are easily dissolved in water. Similarly, the environment around cells is a watery one, the blood and other bodily fluids being solutions in water. Cells then, in order to maintain their integrity, need to be surrounded by an environment through which water cannot flow. A membrane composed of fatty molecules serves this purpose.

We all know from common experience that "oil and water don't mix." That maxim is all one needs to appreciate how a cell is constructed. When oil is poured on water, the oil spreads into a thin film; that film is analogous to the film of fat that surrounds cells, called the plasma membrane. Biological membranes differ from a pure oil film in that the molecules that make the membrane have both oily and watery portions; they have long fatty chains, but they also have a head group that is water-soluble by virtue of being electrically charged. Thus membranes are formed because these bipartite molecules, called phospholipids, spontaneously orient themselves to form a double layer, or bilayer, having a fatty interior with external surfaces bonded to the surrounding water by the charged head groups. The membrane is given rigidity by interspersion of cholesterol, a molecule we have come to hate because of its association with heart disease, but one that is required to build the outer membrane of all our cells. Hence from an understanding of the contrasting properties of watery solutions and oily layers, an understanding of cellular construction emerges.

In spite of the rigidity provided by cholesterol, membranes composed of fat are not very strong, so numerous mechanisms for strengthening the borders of cells have evolved. In plants the plasma membrane is surrounded by a rigid cell wall. Although most animal cells lack a cell wall, proteins attached to their exterior surfaces provide some stability; the linking of cells together through these proteins helps maintain the integrity of tissues. Tissues and organs are often covered by strong networks of proteins and other molecules that strengthen and protect them, and also wall off the various compartments of the body. Single-celled organisms, like bacteria, have special outer coats to protect them.

MEMBRANES SERVE FUNCTIONS OTHER THAN SEGREGATION

Although membranes are valuable as a way to segregate the watery interior of the cell from its environment, or to segregate intracellular events from one another, they have other important functions, including energy storage. Because membranes separate watery compartments from one another, if an ion or a molecule dissolved in water is moved through a membrane into a new cellular compartment, it will not be able to diffuse freely out of the compartment into which it was moved. It takes energy to move the molecule, but once moved, the molecule stores that energy by virtue of its entrapment. Formally, this storage of energy is just like the storage of energy in a battery. Therefore, membranes not only delineate compartments, but also serve as active participants in the cell's dynamism.

The functions of many proteins depend on their mode of association with membranes. For instance, the passage of water-soluble molecules through membranes is carried out by protein transporters that are embedded in the membrane. Also, cells send information to one another by releasing signaling molecules. The outer membranes of cells have proteins, known appropriately as *receptors,* that bind the circulating signaling molecules. These signaling molecules allow the individual activities of the many cells in the body to be coordinated. The receipt of a signaling molecule by a receptor causes the transient organization of particular types of intracellular proteins, called *signal-transduction proteins,* into an activated complex at the interior face of the cell's outer membrane, from which it directs alterations of events in the cell's cytoplasm or nucleus.

PROKARYOTES COMPRISE A SINGLE MEMBRANE LIMITED COMPARTMENT

All prokaryotes are single-celled organisms, or protists. The bacterial lineage includes *Escherichia coli,* found in animal intestines and a favourite experimental organism, and the photosynthetic organisms formerly known as *blue-green algae* but better known today as *cyanobacteria.* (Because most prokaryotes studied in laboratories are bacteria, discussions of prokaryotic structure or metabolism throughout this book refer to these organisms, not archaeans, unless noted otherwise.) Many members of the archaeal lineage grow in unusual, often extreme, environments. For instance, the halophiles require high concentrations of salt to survive, and the thermoacidophiles grow in hot (80°C) sulfur springs, where a pH of less than 2 is common. Other archaeans, called *methanogens,* live in oxygen-free milieus and generate methane (CH_4) by the reduction of carbon dioxide.

Archaeal cells have a similar structure. In general, prokaryotes consist of a single closed compartment containing the cytosol and bounded by the plasma membrane. Although bacterial cells do not have a defined nucleus, the genetic material, DNA, is condensed into the central region of the cell. In addition, most ribosomes — the cell's protein-synthesizing particles — are found in the DNA-free region of the cell. Some bacteria also have an invagination of the cell membrane, called a *mesosome,* which is associated with synthesis of DNA and secretion of proteins. Thus bacterial cells are not completely devoid of internal organization.

Bacterial cells possess a cell wall, which lies adjacent to the external side of the plasma membrane. The cell wall is composed of layers of peptidoglycan, a complex of proteins and oligosaccharides; it helps protect the cell and maintain its shape. Some bacteria (e.g., *E. coli*) have a thin cell wall and an unusual outer membrane separated from the cell wall by the periplasmic space. Such bacteria are not stained by the Gram technique and thus are classified as gram-negative. Other bacteria (e.g., *Bacillus polymyxa*) that have a thicker cell wall and no outer membrane take the Gram stain and thus are classified as gram-positive.

EUKARYOTIC CELLS CONTAIN MANY ORGANELLES AND A COMPLEX CYTOSKELETON

Eukaryotes comprise all members of the plant and animal kingdoms, including the unicellular fungi (e.g., yeasts, mushrooms, molds) and protozoans. Eukaryotic cells, like prokaryotic cells, are surrounded by a plasma membrane. However, unlike prokaryotic cells, most eukaryotic cells also contain extensive internal membranes that enclose specific compartments, the organelles, and separate them from the rest of the cytoplasm, the region of the cell lying outside the nucleus.

Most organelles are surrounded by a single phospholipid membrane, but several, including the nucleus, are enclosed by two membranes. Each type of organelle plays a unique role in the growth and metabolism of the cell, and each contains a collection of specific enzymes that catalyze requisite chemical reactions. The membranes defining these subcellular compartments control their internal ionic composition so that it commonly differs from that of the cytosol (the portion of the cytoplasm outside the organelles) and among the various organelles.

The largest organelle in a eukaryotic cell is generally the nucleus, which houses most of the cellular DNA. In addition to the nucleus, several other organelles are present in nearly all eukaryotic cells: the mitochondria, in which much of the cell's energy metabolism is carried out; the rough and smooth endoplasmic reticula, a network of membranes in which glycoproteins and lipids are synthesized; Golgi vesicles, which direct membrane constituents to appropriate places in the cell; and peroxisomes, in which fatty acids and amino acids are degraded. Animal cells, but not plant cells, contain lysosomes, which degrade worn-out cell constituents and foreign materials taken in by the cell. Chloroplasts, where photosynthesis occurs, are found only in certain leaf cells of plants and some single-celled organisms. Both plant cells and some single-celled eukaryotes contain one or more vacuoles, large, fluid-filled organelles in which nutrients and waste compounds are stored and some degradative reactions occur.

The cytosol of eukaryotic cells contains an array of fibrous proteins collectively called the cytoskeleton. Three classes of fibers compose the cytoskeleton: microtubules (20 nm in diameter), built of polymers of the protein tubulin; microfilaments (7 nm in diameter), built of the protein actin; and intermediate filaments (10 nm in diameter), built of one or more rod-shaped protein subunits. The cytoskeleton gives the cell strength and rigidity, thereby helping to maintain cell shape. Cytoskeletal fibers also control movement of structures within the cell; for example, some cytoskeletal fibers connect to organelles or provide tracks along which organelles move.

The rigid cell wall, composed of cellulose and other polymers, that surrounds plant cells contributes to their strength and rigidity. Fungi are also surrounded by a cell wall, but its composition differs from that of bacterial or plant cell walls.

CELLULAR DNA IS PACKAGED WITHIN CHROMOSOMES

The DNA in the nuclei of eukaryotic cells is distributed among 1 to more than 50 long linear structures called chromosomes. The number and size of the chromosomes are the same in all cells of an organism, but vary among different types of organisms. Each chromosome comprises a single DNA molecule associated with numerous proteins, and the total DNA in the chromosomes of an organism is referred to as its genome. Chromosomes, which stain intensely with basic dyes, are visible in the light microscope only during cell division when the DNA becomes tightly compacted.

In all prokaryotic cells, most of or all the genetic information resides in a single circular DNA molecule, about a millimeter in length; this molecule lies, folded back on itself many times, in the central region of the cell. Although the large genomic DNA molecule in prokaryotes is associated with proteins and often is referred to as a chromosome, the arrangement of DNA within a bacterial chromosome differs greatly from that within the chromosomes of eukaryotic cells.

The concept that genes are like "beads" strung on a long "string," the chromosome, was proposed early in the 1900s based on genetic work with the fruit fly *Drosophila*. The early *Drosophila* workers could position, or map, the genes responsible for various mutant traits on a chromosome, even though they did not yet know that genes were segments of DNA or that the function of a gene was due to a protein whose sequence was encoded by that gene!

THE LIFE CYCLE OF CELLS

A cell in an adult organism can be viewed as a steady-state system. The DNA is constantly read out into a particular set of mRNAs, which specify a particular set of proteins. As these proteins function, they are also being degraded and replaced by new ones, and the system is so balanced that the cell neither grows, shrinks, nor changes its function. This static view of the cell, however, misses the all-important dynamic aspects of cellular life.

The dynamics of a cell can best be understood by examining the course of its life. A new cell arises when one cell divides or when two cells, like a sperm and an egg cell, fuse. Either event sets off a cell-replication programme that is encoded in the DNA and executed by proteins. This programme usually involves a period of cell growth, during which proteins are made and DNA is replicated, followed by cell division, when a cell divides into two daughter cells. Whether a given cell will grow and divide is a highly regulated decision of the body, assuring that an adult organism replaces worn out cells or makes more cells in response to a new need. Examples of the latter are the growth of muscle in response to exercise or damage, and the proliferation of red blood cells when a person ascends to a higher altitude and needs more capacity to capture oxygen. However, in one major and devastating disease — cancer — cells multiply even though they are not needed by the body. To understand how cells become cancerous, biologists have intensely studied the mechanisms that control the growth and division of cells.

THE CELL CYCLE FOLLOWS A REGULAR TIMING MECHANISM

Most eukaryotic cells live according to an internal clock; that is, they proceed through a sequence of phases, called the cell cycle, during which DNA is duplicated during the synthesis (S) phase and the copies are distributed to opposite ends of the cell during mitotic (M) phase. Progress along the cycle is controlled at key checkpoints, which monitor the status of a cell, for instance, the internal amount of DNA or the presence of extracellular nutrients. When certain conditions are met, the cell proceeds to the next checkpoint. The cycle begins after the cell divides into two daughter cells, each containing an identical copy of the parental cell's genetic material.

The cell cycle of prokaryotes is simple and fast. Replication of the single chromosome begins at a particular DNA sequence, the replication origin, which is anchored to the cell membrane. Once DNA replication is complete, assembly of new membrane and cell wall forms a septum, which eventually divides the cell in two. Because the origins of the two newly formed chromosomes are anchored to different membrane sites, each daughter cell receives one chromosome. In ideal growth conditions, the bacterial cell cycle is repeated every 30 minutes.

Only a few types of eukaryotic cells can grow and divide

as quickly as bacteria. Most growing plant and animal cells take 10 – 20 hours to double in number, and some duplicate at a much slower rate. Many cells in adult animals, such as nerve cells and striated muscle cells, do not divide at all. They have temporarily exited from the cell cycle after mitosis and entered a "paused or quiescent" state called G_0. Because eukaryotic cells are larger and more complex than prokaryotic cells, a specialized mechanism coordinates their replication of genomic DNA, distribution of chromosomes, and cell division.

MITOSIS APPORTIONS

Mitosis is the mechanism in eukaryotes for partitioning the genome equally at cell division. To accomplish this complex task, plant and animal cells build a specialized machine, called the mitotic apparatus, which captures the chromosomes and then pushes and pulls them to opposite sides of the dividing cell. Remarkably, the mitotic apparatus is a temporary structure that exists only during mitosis to distribute the genetic material.

Although the events of mitosis unfold continuously, they are conventionally divided into four substages representing phases of chromosome movement. During the first substage, prophase, the replicated chromosomes, each comprising two identical chromatids, are condensed into compact packets and then released to the cytoplasm when the nuclear membrane breaks down.

During metaphase and anaphase, the chromosomes are sorted, and each chromatid of a pair moves to opposite sides of the cell. The end of mitosis is marked by re-formation of a membrane around each set of chromosomes (telophase). Division of the cytoplasm, called cytokinesis, then yields two daughter cells, each with a $2n$ complement of genetic material. Cell division in plant and animal cells differs mainly at cytokinesis. Animal cells divide in two by pinching of the cytoplasm. However, because a plant cell is surrounded by a rigid cell wall, daughter cells are formed by building a new cell membrane and cell wall between the two daughter nuclei, thereby cutting the cytoplasm into two portions.

CELL DIFFERENTIATION

The most complicated example of cellular dynamics occurs when a cell changes, or *differentiates,* to carry out a specialized function. This process often is marked by a change in the microscopic appearance, or *morphology,* of the cell. For example, the different structures of a nerve cell and a muscle cell reflect their respective functions in long-distance communication and contraction, highlighting the biological principle that "form follows function."

Cell differentiation creates the diversity of cell types that arise during the development of an organism from a fertilized egg. This is a process of extensive cell multiplication and differentiation. A mammal that starts as one cell becomes an organism with hundreds of diverse cell types such as muscle, nerve, and skin. Here we see at its most dramatic the power of DNA to control cellular behaviour: development is a DNA-orchestrated set of cellular changes (easily tens of thousands of them) that occur virtually without fail. The almost perfect resemblance of "identical" twins is a testament to the programme encoded by DNA to reproducibly direct the development of a human being.

Nowhere is the variety of cellular activities and responses better illustrated than in the body's immune system. It is there that many cell types come together in organized tissues specifically designed to allow the body to distinguish its own cells from those of foreign invaders. Within the immune system, we see both development of specialized cells that can recognize invading cells and formation of tissues from cells that originate in various parts of the body. The immune-system cells not only actively survey their environment with surface receptor proteins like antibodies, but also change their properties when they encounter a foreign substance, allowing the body to rid itself of invaders.

CELLS DIE BY SUICIDE

Unchecked cell growth and multiplication produce a mass of cells, a tumor. *Programmemed cell death* plays the very important role of population control by balancing cell growth and multiplication. In addition, cell death also eliminates unnecessary cells. For example, during embryogenesis, the digits of our fingers and toes are sculpted by the death of cells in the intervening spaces. If these cells remained alive, our hands and feet would become webbed. Thus the timing and location of cell death, as well as cell growth and division, must be precisely controlled.

Cell death follows an internal programme of events called apoptosis, in which all traces of a cell vanish. The first visible sign of apoptosis is condensation of the nucleus and fragmentation of the DNA. The cell soon shrivels and is consumed by macrophages. A cell is directed to commit suicide when an essential factor is removed from the extracellular environment or when an internal signal is activated. Thus, the default state of the cell is to remain alive. The discovery of genes that suppress the growth of tumors by activating cell death stimulated an exciting new line of cancer research that may lead to more effective treatment strategies

INTEGRATING CELLS INTO TISSUES

The evolution of multicellular organisms permitted specialized cells and tissues to form; a flowering plant has at least 15 cell types, and a vertebrate hundreds. In both plants and animals, cells that are specialized to carry out a particular task are found together in the tissues in which the task is performed: a xylem or meristem; a liver, a muscle, or a nerve ganglion. Different types of cells in a tissue are often arranged in precise patterns of staggering complexity. For instance, the hundreds of different types of neurons in the human brain are interconnected to one another through a network of some 10^{15} synaptic connections! The coordinated functioning of many types of cells within tissues, and of multiple specialized tissues, permits the organism as a whole to move, metabolize, reproduce, and carry out other essential activities.

A key step in the evolution of multicellularity must have been the ability of cells to contact tightly and interact specifically with other cells. Various integral membrane proteins, collectively termed cell-adhesion molecules (CAMs), enable many animal cells to adhere tightly and specifically with cells of the same, or similar, type; these interactions allow populations of cells to segregate into distinct tissues. Following aggregation, cells elaborate specialized cell junctions that stabilize these interactions and promote local communication between adjacent cells.

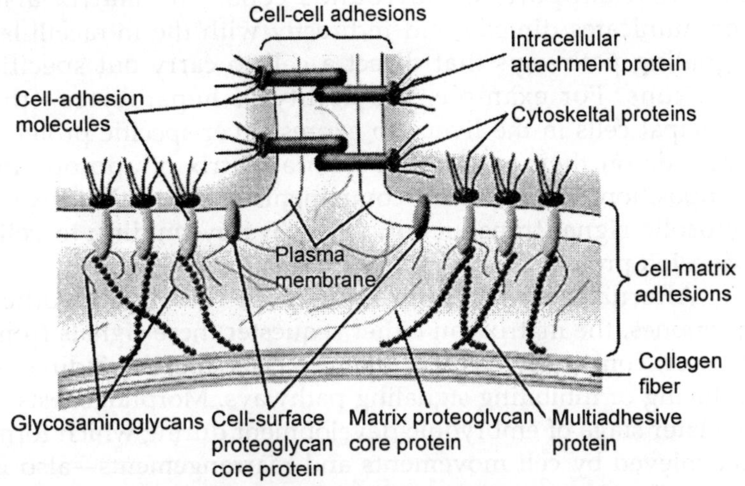

Fig. Integrating Cells into Tissues.

Animal cells also secrete a complex network of proteins and carbohydrates, the extracellular matrix (ECM), that creates a special environment in the spaces between cells. The matrix helps bind the cells in tissues together and is a reservoir for many hormones controlling cell growth and differentiation. The matrix also provides a lattice through which cells can move, particularly during the early stages of differentiation. Defects in these connections lead to cancer and developmental malformations.

The extracellular matrix has three major protein components: highly viscous proteoglycans, which cushion cells; insoluble collagen fibers, which provide strength and resilience; and soluble multiadhesive matrix proteins, which bind these components to receptors on the cell surface. Different combinations of these components tailor the strength of the extracellular matrix for different purposes. For example, animals contain many types of extracellular matrices, each specialized for a particular function such as strength (in a tendon), cushioning (in cartilage), or adhesion. In the case of smooth muscle cells that surround an artery, the extracellular matrix must provide strong but flexible connections.

The extracellular matrix is not just an inert framework or cage that supports or surrounds cells. The matrix also communicates directly and indirectly with the intracellular signaling pathways that direct a cell to carry out specific functions. For example, the ability of hepatocytes—the principal cells in the liver—to express liver-specific proteins depends on their association with a matrix of appropriate composition. Specific ECM components can directly activate cytosolic signal-transduction pathways by binding to cell-adhesion protein receptors in the plasma membrane.

Alternatively, by binding growth factors and other hormones, the matrix can either sequester these signals from cells or, conversely, present them to cells, thereby indirectly inducing or inhibiting signaling pathways. Morphogenesis—the later stage of embryonic development during which form is achieved by cell movements and rearrangements—also is critically dependent on ECM components, which are constantly being remodeled, degraded, and resynthesized locally. Even in adults—in areas of wounding, for example degradation and resynthesis of ECM components occurs.

In this chapter, we focus on the structure of and interactions between ECM components, cell-adhesion molecules, and cell-adhesion junctions—the main structures that permit animal cells to form organized tissues. Plant cells are surrounded by a cell wall that is thicker and more rigid than the extracellular matrix. Although the plant cell wall and the extracellular matrix serve many of the same functions, they are structurally very different.

Chapter 10

The Eukaryotic Cells

Recognition that modern-day organisms fall into three distinct evolutionary lineages came from comparisons of the nucleotide sequences of genes common to all organisms and of ribosomal RNAs, particularly the RNA species found in the small ribosomal subunit. Prokaryotes, which lack a defined nucleus and have a simple subcellular organization, form two of the lineages — the bacteria and archaea. The third lineage is the eukaryotes (eukarya), whose cells have a membrane-limited nucleus containing most of the cellular DNA, numerous specialized organelles, and a complex cytoskeleton. Despite the differences in cellular organization between prokaryotes and eukaryotes, all cells share certain structural features and carry out DNA replication, protein synthesis, production of ATP, and many other complicated metabolic events in basically the same way. At the molecular level there is a close relationship between protein structure and function. Likewise, at the cellular level, function and structure are intimately related. Thus this chapter describes the basic structural components of cells and the general functions of each, focusing on eukaryotic cells.

We begin by describing the capabilities and applications of various techniques of microscopy used to visualize cells and subcellular structures. We then consider the basic methods for purifying biological membranes and subcellular structures. The purified preparations produced by cell fractionation permit detailed studies on the structure and function of biomembranes and other subcellular structures. Following this overview of experimental techniques, we continue the discussion of biomembranes begun in earlier chapters. The surface membrane found on all cells and the membranes that line eukaryotic organelles all have the same basic architecture — a phospholipid bilayer; the unique function of each type of membrane is determined primarily by the proteins it contains. The final section briefly describes the structure and function of the main internal organelles and fibers of eukaryotic cells, as well as the extracellular substances that surround cells and give them shape and strength. This review sets the stage for later discussions of cell function at the molecular level.

MICROSCOPY AND CELL ARCHITECTURE

The modern, detailed understanding of cell architecture is based on several types of microscopy. Because there is no one "correct" view of a cell, it is essential to understand the characteristics of the key cell-viewing techniques, the types of images they produce, and their limitations.

Schleiden and Schwann, using a primitive light microscope, first described individual cells as the fundamental unit of life, and light microscopy has continued to play a major role in biological research. The development of electron microscopes greatly extended the ability to resolve subcellular particles and has yielded much new information on the organization of plant and animal tissues. The nature of the images depends on the type of light or electron microscope employed and on the way in which the cell or tissue has been prepared. Each technique is designed to emphasize particular structural features of the cell. The epithelial cell lining the small intestine, appears when viewed by three different microscopic techniques.

In this section, we focus on the most common application of light and electron microscopy — to visualize fixed, killed cells. Although this approach reveals much information, a critical question about such results is how true to life is the image of a biological specimen that has been fixed, stained, and dehydrated before examination? Thus we also consider some of the refinements that allow microscopy of unaltered or less altered specimens.

Light Microscopy

The *compound microscope,* the most common microscope in use today, contains several lenses that magnify the image of a specimen under study. The total magnification is a product of the magnification of the individual lenses: if the *objective lens* magnifies 100-fold (a 100X lens, the maximum usually employed) and the *eyepiece* magnifies 10-fold, the final magnification recorded by the human eye or on film will be 1000-fold.

However, the most important property of any microscope is not its magnification but its resolving power, or resolution — its ability to distinguish between two very closely positioned objects. Merely enlarging the image of a specimen accomplishes nothing if the image is blurry. The resolution of a microscope lens is numerically equivalent to *D,* the minimum distance between two distinguishable objects; the smaller the value of *D,* the better the resolution. *D* depends on three parameters, all of which must be considered in order to achieve the best possible resolution: the *angular aperture,* a, or half-angle of the cone of light entering the objective lens from the specimen; the *refractive index, N,* of the air or fluid medium between the specimen and the objective lens; and the *wavelength,* l, of incident light: $D = (0.61l) \div (N \times \sin a)$. Decreasing the value of l or increasing either *N* or a will decrease the value of *D* and thus improve the resolution. Note that the magnification is not part of this equation.

The angular aperture, a, depends on the width of the objective lens and its distance from the specimen. Moving the objective lens closer to the specimen increases the angle a and thus sin a, and therefore reduces *D* (i.e., increases the resolution).

Intuitively, one can recognize that increasing a allows a greater fraction of the light emanating from the specimen to enter the objective lens. The refractive index *N* is a measure of the degree to which a medium bends a light ray that passes through it; the refractive index of air is defined as 1.0. Use of immersion oil, which has a refractive index of 1.5, is a simple way to reduce *D* by 33 percent. An intuitive explanation for this improvement is that a medium with a higher refractive index than air, if placed between the specimen and the objective lens, will "bend" more of the light emanating from the specimen such that it goes into the lens. Finally, the shorter the wavelength of incident light, the lower will be the value of D and the better the resolution.

Due to limitations on the values of α, λ, and N, the *limit of resolution* of a light microscope using visible light is about 0.2 μm (200 nm). No matter how many times the image is magnified, the microscope can never resolve objects that are less than ≈0.2 μm apart or reveal details smaller than ≈0.2 μm in size. This is true because the maximum angular aperture for the best objective lenses is 70° (sin 70° = 0.94). With the visible light of shortest wavelength (blue, $\lambda = 450$ nm) and with an immersion oil ($N = 1.5$) above the sample, then or about 0.2 μm.

$$D = \frac{0.61 \times 450nm}{1.5 \times 0.94} = 194nm$$

Despite this limit of resolution, the light microscope can be used to track the location of a small bead of known size to a precision of only a few nanometers! If we know the precise size and shape of an object — say, a 5-nm sphere of gold — and if we use a video camera to record the microscopic image as a digital image, then a computer can calculate the position of the *centre* of the object to within a few nanometers. This technique has been used, to nanometer resolution, for tracking the movement of gold particles attached via antibodies to specific proteins on the surface of living cells.

SAMPLES FOR LIGHT MICROSCOPY USUALLY ARE FIXED, SECTIONED, AND STAINED

Specimens for light microscopy are commonly fixed with a solution containing alcohol or formaldehyde, compounds that denature most proteins and nucleic acids. Formaldehyde also cross-links amino groups on adjacent molecules; these covalent bonds stabilize protein-protein and protein – nucleic acid interactions and render the molecules insoluble and stable for subsequent procedures. Usually the sample is then embedded in paraffin or plastic and cut into thin sections of one or a few micrometers thick. Alternatively, the sample can be frozen without prior fixation and then sectioned; this avoids the denaturation of enzymes by fixatives such as formaldehyde.

Since the resolution of the light microscope is 0.2 mm and mitochondria and chloroplasts are 1 mm long (about the size of bacteria), theoretically one should be able to see these organelles. However, most cellular constituents are not coloured and absorb about the same degree of visible light, so that they are hard to distinguish under a light microscope unless the specimen is stained. Thus the final step in preparing a specimen for light microscopy is to stain it, in order to visualize the main structural features of the cell or tissue.

Many chemical stains bind to molecules that have specific features. For example, *hematoxylin* binds to basic amino acids (lysine and arginine) on many different kinds of proteins, whereas *eosin* binds to acidic molecules (such as DNA, and aspartate and glutamate side chains). Because of their different binding properties, these dyes stain various cell types sufficiently differently that they are distinguishable visually. Two other common dyes are *benzidine,* which binds to heme-containing proteins and nucleic acids, and *fuchsin,* which binds to DNA and is used in Fuelgen staining.

If an enzyme catalyzes a reaction that produces a coloured or otherwise visible precipitate from a colourless precursor, the enzyme may be detected in cell sections by their coloured reaction products. This technique is called *cytochemical staining.*

FLUORESCENCE MICROSCOPY CAN LOCALIZE AND QUANTIFY SPECIFIC MOLECULES IN CELLS

Perhaps the most versatile and powerful technique for localizing proteins within a cell by light microscopy is fluorescent staining of cells and observation in the *fluorescence microscope.* A chemical is said to be *fluorescent* if it absorbs light at one wavelength (the *excitation wavelength*) and emits light (fluoresces) at a specific and longer wavelength. Most fluorescent dyes emit visible light, but some (such as Cy5 and

Cy7) emit infrared light. In modern fluorescence microscopes, only fluorescent light emitted by the sample is used to form an image; light of the exciting wavelength induces the fluorescence but is then not allowed to pass the filters placed between the objective lens and the eye or camera.

REVEALING SPECIFIC PROTEINS IN FIXED CELLS

Four very useful dyes for fluorescent staining are rhodamine and Texas red, which emit red light; Cy3, which emits orange light; and fluorescein, which emits green light. These dyes have a low, nonspecific affinity for biological molecules, but they can be chemically coupled to purified antibodies specific for almost any desired macromolecule. When a fluorescent dye – antibody complex is added to a permeabilized cell or tissue section, the complex will bind to the corresponding antigens, which then light up when illuminated by the exciting wavelength, a technique called *immunofluorescence microscopy*. By staining a specimen with two or three dyes that fluoresce at different wavelengths, multiple proteins can be localized within a cell.

REVEALING SPECIFIC PROTEINS IN LIVING CELLS

Fluorescence microscopy can also be applied to live cells. For example, purified actin may be chemically linked to a fluorescent dye. Careful biochemical studies have established that this "tagged" molecule is indistinguishable in function from its normal counterpart. If the tagged protein is *microinjected* into a cultured cell, the endogenous cellular and injected tagged actin monomers copolymerize into normal long actin fibers. This technique can also be used to study individual microtubules within a cell.

Another technique for detecting specific proteins within living cells takes advantage of *green fluorescent protein* (GFP), a naturally fluorescent protein found in the jellyfish *Aequorea victoria*. The bioluminescence of this organism, which radiates a green fluorescence, is due to GFP. This 238-aa protein contains serine, tyrosine, and glycine residues whose side chains have spontaneously reacted with one another to form a fluorescent chromophore. By recombinant DNA techniques the GFP gene can be introduced into living cultured cells or into specific cells of an entire animal. Because the introduced gene will express GFP, the cells will emit a green fluorescence when irradiated; this GFP fluorescence can be used to localize the cells within a tissue.

Alternatively, the gene for GFP can be fused to the gene for another protein of interest, producing a recombinant DNA encoding one long chimeric protein that contains the entirety of both proteins. Cells in which this recombinant DNA has been introduced will synthesize this chimeric protein, whose green fluorescence will reveal the subcellular localization of the protein.

DETERMINING THE INTRACELLULAR CONCENTRATION OF CA^{2+} AND H^{+} IONS

Changes in the cytosolic concentration of Ca^{2+} ions or pH frequently signal changes in cellular metabolism. The Ca^{2+} concentration in the cytosol of resting cells, for instance, is about 10^{-7} M. Many hormones or other stimuli cause a rise in cytosolic Ca^{2+} to 10^{-6} M; this, in turn, causes changes in cellular metabolism, such as contraction of muscle.

The fluorescent properties of certain dyes, such as *fura-2*, facilitate measurement of the concentration of free Ca^{2+} in the cytosol. This dye contains five carboxylate groups that form ester linkages with ethanol. The resulting fura-2 ester is lipophilic and can diffuse from the medium across the plasma membrane into cells. Within the cytosol, esterases hydrolyze fura-2 ester yielding fura-2, whose free carboxylate groups render the molecule nonlipophilic, so it cannot cross cellular membranes and remains in the cytosol.

Each fura-2 molecule can bind a single Ca^{2+} ion but no other cellular cation, and the amount of fura-2 bound to Ca^{2+} is proportional, over a certain range, to the Ca^{2+} concentration. The fluorescence of fura-2 at one particular wavelength is enhanced when Ca^{2+} is bound, and the fluorescence is proportional to the Ca^{2+} concentration. At another wavelength the fluorescence of fura-2 is the same whether or not Ca^{2+} is bound and provides a measure of the total amount of fura-2 in the segment of the cell. By examining cells continuously in the fluorescence microscope and measuring rapid changes in the ratio of fura-2 fluorescence at these two wavelengths, one can quantify rapid changes in the fraction of fura-2 that has a bound Ca^{2+} ion and thus in the concentration of cytosolic Ca^{2+}.

The fluorescence of other dyes is sensitive to the H^+ concentration and can be used in a similar way to monitor the cytosolic pH of living cells.

CONFOCAL SCANNING

Immunofluorescence microscopy has its limitations. The fixatives employed to preserve cell architecture often destroy the *antigenicity* of a protein, that is, its ability to bind to its specific antibody. Also, the method generally gives poor results with thin cell sections, because embedding media often fluoresce themselves, obscuring the specific signal from the antibody.

Moreover, in microscopy of whole cells, the fluorescent light comes from molecules above and below the plane of focus; thus the observer sees a superposition of fluorescent images from molecules at many depths in the cell, making it difficult to determine the actual three-dimensional molecular arrangement.

The *confocal scanning microscope* avoids the last problem by permitting the observer to visualize fluorescent molecules in a single plane of focus, thereby creating a vastly sharper cross-sectional image. At any instant during confocal imaging, only a single small part of a sample is illuminated with exciting light from a focused laser beam, which rapidly moves to different spots in the sample focal plane. Images from these spots are recorded by a video camera and stored in a computer, and the composite image is displayed on a computer screen.

Deconvolution microscopy is similar to confocal microscopy in that a cross-sectional image is obtained, but the two techniques differ in the details of how this image is generated. In both cases, the objective lens collects light that originates from above and below the focal plane as well as that which originates from within the focal plane. Confocal microscopes use a pinhole to exclude the out-of-focus light. In contrast, deconvolution microscopes collect all the light from several focal planes, and then mathematically reassign the out-of-focus light to its correct focal plane with the aid of a high-speed computer, a mathematical operation called *deconvolution*.

To understand how a deconvolution microscope works, consider an infinitely small fluorescent source of light, which can be approximated by a fluorescent bead smaller than the resolution of the light microscope (i.e., <0.2 mm in diameter). The emitted light radiates in all directions, and when the source is in the focal plane of the objective, it appearsas a bright point of light.

When the point source is outside the focal plane of the objective, some of the light is still collected by the objective lens, and the point source appears as a halo. As the focal plane is moved farther away from the plane containing the point source, the halo becomes larger and more diffuse.

Knowing exactly how the light emitted by an infinitely small fluorescent source is collected and distorted by the optics of the sample and microscope, it is possible to reconstruct an individual cross-sectional image (containing only light that originated in the focal plane of interest) from a set of images taken as the objective focal plane is moved through the plane of interest.

Cross-sectional images obtained with a deconvolution microscope may have even greater detail than those obtained with a confocal microscope. Additionally, the fluorescent labeling of the sample does not need to be as intense for deconvolution microscopy as it does for confocal microscopy, since all the light produced by a fluorescent sample is collected and analyzed by the microscope.

Three-dimensional images can be obtained by a refinement known as *optical sectioning*. In this method, a computer records individual fluorescent images of planes at different depths of the sample — in effect, serial sections — and combines the stack of images into one three-dimensional image.

PHASE-CONTRAST

Detailed views of transparent, live, unstained cells and tissues are obtainable with *phase-contrast microscopy* and *Nomarski interference microscopy*. Both techniques take advantage of the phenomena of refraction and diffraction of light waves. As a result, small differences in refractive index and thickness between parts of the specimen (say, between the nucleus and cytosol) or between the specimen and the surrounding medium can be converted into differences of light and dark in the final image.

The phase-contrast microscope generates an image in which the degree of darkness or brightness of a region of the sample depends on the refractive index of that region. The improved definition of subcellular structures in live, unstained cells obtained by phase-contrast microscopy compared with standard bright-field microscopy.

Nomarski, or differential, interference microscopy generates an image that looks as if the specimen is casting a shadow to one side: the "shadow" primarily represents a difference in refractive index and thickness of a specimen rather than its topography. In this technique, a prism splits an incident beam of plane-polarized light so that one part of the beam passes through one region of a specimen and the other part passes through a closely adjacent region; a second prism then reassembles the two beams. Minute differences in thickness or in the refractive index between adjacent parts of a sample are converted into a bright image (if the two beams are in phase when they recombine) or a dark one (if they are out of phase).

Phase-contrast microscopy is especially useful in examining the structure and movement of larger organelles, such as the nucleus and mitochondria, in live cultured cells. The greatest disadvantage of this technique is that it is suitable for observing only single cells or thin cell layers. Nomarski interference microscopy, in contrast, defines only the outlines of large organelles, such as the nucleus and vacuole. However, thick objects, such as the nuclei in a worm, can be observed by combining this technique with optical sectioning.

Both phase-contrast and Nomarski interference microscopy can be used in *time-lapse microscopy,* in which the same cell is photographed at regular intervals over periods of several hours. This procedure allows the observer to study cell movement, provided the microscope's stage can control the temperature of the specimen and the gas environment.

TRANSMISSION ELECTRON MICROSCOPY HAS A LIMIT OF RESOLUTION OF 0.1 NM

The fundamental principles of electron microscopy are similar to those of light microscopy; the major difference is that electromagnetic lenses, not optical lenses, focus a high velocity electron beam instead of visible light. Because electrons are absorbed by atoms in air, the entire tube between the electron source and the viewing screen is maintained under an ultrahigh vacuum.

The *transmission electron microscope* (TEM) directs a beam of electrons through a specimen. Electrons are emitted by a tungsten cathode when it is electrically heated. The electric potential of the cathode is kept at 50,000 – 100,000 volts; that of the anode, near the top of the tube, is zero. This drop in voltage causes the electrons to accelerate as they move toward the anode. A condenser lens focuses the electron beam onto the sample; objective and projector lenses focus the electrons that pass through the specimen and project them onto a viewing screen or a piece of photographic film.

In typical electron microscopes, electrons have the properties of a wave with a wavelength of only 0.005 nm. Recall that the minimum distance *D* at which two objects can be distinguished is proportional to the wavelength l of the light that illuminates the objects. Thus the limit of resolution for the electron microscope is theoretically 0.005 nm (less than the diameter of a single atom), or 40,000 times better than the resolution of the light microscope and 2 million times better than that of the unaided human eye. However, the effective resolution of the electron microscope in the study of biological systems is considerably less than this ideal. Under optimal conditions, a resolution of 0.10 nm can be obtained with transmission electron microscopes, about 2000 times better than the best resolution of light microscopes.

PREPARATION OF FIXED, STAINED SAMPLES FOR TEM

Like the light microscope, the transmission electron microscope is used to view thin sections of a specimen, but the fixed sections must be much thinner for electron microscopy (only 50 – 100 nm, about 0.2 percent of the thickness of a single cell). Clearly, only a small portion of a cell can be observed in any one section. Generation of the image depends on differential scattering of the incident electrons by molecules in the preparation. Without staining, the beam of electrons passes through a cell or tissue sample uniformly, so the entire sample appears uniformly bright with little differentiation of components. Staining techniques are therefore used to reveal the location and distribution of specific materials.

Heavy metals, such as gold or osmium, appear dark on a micrograph because they scatter (diffract) most of the incident electrons; scattered electrons are not focused by the electromagnetic lenses and do not form the image. Osmium tetrox-ide preferentially stains certain cellular components, such as membranes, which appear black in micrographs. Specific proteins can be detected in thin sections by use of electron-dense gold particles coated with protein A, a bacterial protein that binds antibody molecules nonspecifically.

Electron microscopy also is used to obtain information about the shapes of purified viruses, fibers, enzymes, and other subcellular particles. In one technique, called *metal shadowing,* a thin layer of evaporated metal, such as platinum, is laid at an angle on a biological sample. An acid bath dissolves the biological material, leaving a metal replica of its surface, which can then be examined in the transmission electron microscope. Variations in the angle and thickness of the deposited metal allow an image to be formed because some incident electrons will be scattered in various directions rather than pass through the preparation. If the metal is deposited mainly on one side of the sample, for instance, the image seems to have "shadows," where the metal appears dark and the shadows appear light.

CRYOELECTRON MICROSCOPY

Standard electron microscopy cannot be used to study live cells because they are generally too vulnerable to the required conditions and preparatory techniques. In particular, the absence of water causes macromolecules to become denatured and nonfunctional. However, the technique of *cryoelectron microscopy* allows examination of hydrated, unfixed, and unstained biological specimens directly in the transmission electron microscope.

In this technique, an aqueous suspension of a sample is applied in an extremely thin film to a grid. After it has been frozen in liquid nitrogen and maintained in this state by means of a special mount, it is observed in the electron microscope. The very low temperature ("196 °C) keeps the water from evaporating, even in a vacuum, and the sample can be observed in detail in its native, hydrated state without shadowing or fixing it. By computer-based averaging of images of hundreds of particles, a three-dimensional model almost to atomic resolution can be generated.

SCANNING ELECTRON MICROSCOPY

The *scanning electron microscope* allows the investigator to view the surfaces of unsectioned specimens. These cannot be visualized with transmission equipment because the electrons pass through the entire specimen. The sample is fixed, dried, and coated with a thin layer of a heavy metal, such as platinum, by evaporation in a vacuum; in this case, the sample is rotated so that the platinum is deposited uniformly on the surface. An intense electron beam inside the microscope scans rapidly over the sample. Molecules in the specimen are excited and release secondary electrons that are focused onto a scintillation detector; the resulting signal is displayed on a cathode-ray tube. Because the number of secondary electrons produced by any one point on the sample depends on the angle of the electron beam in relation to the surface, the scanning electron micrograph has a threedimensional appearance. The resolving power of scanning electron microscopes, which is limited by the thickness of the metal coating, is only about 10 nm, much less than that of transmission instruments.

PURIFICATION OF CELLS AND THEIR PARTS

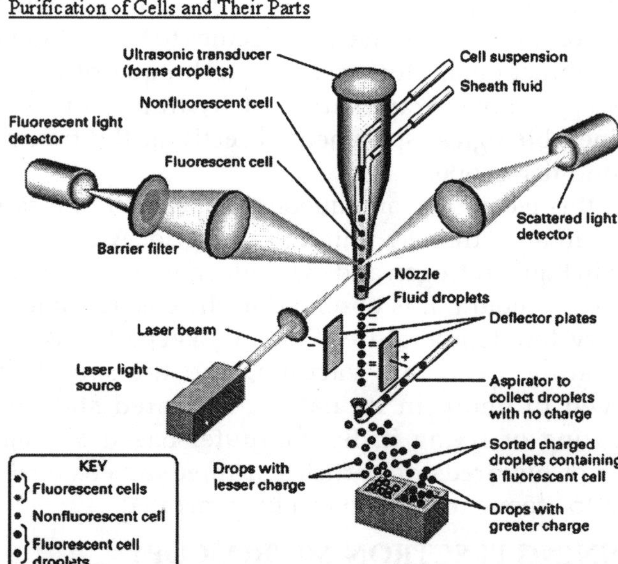

Fig. Fluorescence-activated cell sorter (FACS)

Most animal and plant tissues contain a mixture of cell types. However, an investigator often wishes to study a pure population of one type of cell. In some cases, cells differ in some physical property that allows different cell types to be separated. White blood cells (leukocytes) and red blood cells (erythrocytes), for instance, have very different densities because erythrocytes have no nucleus; thus these cells can be separated on the basis of density.

Since most cell types cannot be differentiated so easily, other cell-separation techniques have had to be developed. Similarly, it is essential to isolate quantities of each of the major subcellular organelles to study their structures and metabolic functions in detail.

FLOW CYTOMETRY SEPARATES DIFFERENT CELL TYPES

A *flow cytometer* can identify different cells by measuring the light they scatter, or the fluorescence they emit, as they flow through a laser beam; thus it can sort out cells of a particular type from a mixture. Indeed, a *fluorescence-activated cell sorter (FACS),* an instrument based on flow cytometry, can select one cell from thousands of other cells. For example, if an antibody specific to a certain cell-surface molecule is linked to a fluorescent dye, any cell bearing this molecule will bind the antibody and will then be separated from other cells when it fluoresces in the FACS. Once sorted from the other cells, the selected cell can be grown in culture.

This procedure is commonly used to purify the different types of white blood cells, each of which bears on its surface one or more distinctive proteins and thus will bind monoclonal antibodies specific for that protein. Such FACS separations are more difficult to conduct on cultured cells or cells from animal tissues, which interact with adjacent cells and are surrounded by an extracellular matrix. Samples must be treated with proteases to degrade the extracellular-matrix proteins and cell-surface proteins that attach cells in tissues to one another; these proteases usually also degrade the distinctive cell-surface "marker" proteins that distinguish one cell type from another.

Other uses of flow cytometry include the measurement of a cell's DNA and RNA content and the determination of its general shape and size. The FACS can make simultaneous measurements of the size of a cell (from the amount of scattered light) and the amount of DNA it contains (from the amount of fluorescence from a DNA-binding dye).

DISRUPTION OF CELLS

The initial step in purifying subcellular structures is to rupture the plasma membrane and the cell wall, if present. First, the cells are suspended in a solution of appropriate pH and salt content, usually isotonic sucrose (0.25 M) or a combination of salts similar in composition to those in the cell's interior. Many cells can then be broken by stirring the cell suspension in a high-speed blender or by exposing it to highfrequency sound *(sonication)*. Plasma membranes can also be sheared by special pressurized tissue homogenizers in which the cells are forced through a very narrow space between the plunger and the vessel wall. Generally, the cell solution is kept at 0 °C to best preserve enzymes and other constituents after their release from the stabilizing forces of the cell.

Because the plasma membrane is highly permeable to water but poorly permeable to the salts and other small molecules (solutes) within cells, *osmotic flow* can be enlisted to help rupture cells. Recall that water flows across a semipermeable membrane, such as the plasma membrane, from a solution of high water (low solute) concentration to one of low water (high solute) concentration until the water concentration on both sides is equal.

Consequently, when cells are placed in a hypotonic solution (i.e., one with a lower salt concentration than that of the cell interior), water flows into the cells. This osmotic flow causes the cells to swell and then more easily rupture. Conversely, in a hypertonic solution (i.e., one with a higher salt concentration than that of the cell interior), water flows out of cells, causing them to shrink. When cells are placed in an isotonic solution (i.e., one with a salt concentration equal to that of the cell interior), there is no net movement of water in or out of cells. For this reason, an isotonic solution is best for preserving normal cell structure.

Disrupting the cell produces a mix of suspended cellular components, the *homogenate,* from which the desired organelles can be retrieved. Because rat liver contains an abundance of a single cell type, this tissue has been used in many classic studies of cell organelles. However, the same isolation principles apply to virtually all cells and tissues, and modifications of these cell-fractionation techniques can be used to separate and purify any desired components.

DIFFERENT ORGANELLES CAN BE SEPARATED BY CENTRIFUGATION

We discussed the principles of centrifugation and the uses of centrifugation techniques for separating proteins and nucleic acids. Similar approaches are used for separating and purifying the various organelles, which differ in both size and density. Most fractionation procedures begin with *differential centrifugation* at increasingly higher speeds also called *differential-velocity centrifugation.* The different sedimentation rates of various cellular components make it possible to separate them partially by centrifugation. Nuclei and viral particles can sometimes be purified completely by such a procedure. After centrifugation at each speed for an appropriate time, the supernatant is poured off and centrifuged at higher speed. Each pelleted fraction can be resuspended and further separated by equilibrium densitygradient centrifugation (discussed next).

Differential centrifugation does not yield totally pure organelle fractions. One method for further purifying fractions is *equilibrium density-gradient centrifugation,* which separates cellular components according to their density. The impure organelle fraction is layered on top of a solution that contains a gradient of a dense nonionic substance, such as sucrose or glycerol. The tube is centrifuged at a high speed (about 40,000 rpm) for several hours, allowing each particle to migrate to an equilibrium position where the density of the surrounding liquid is equal to the density of the particle. In typical preparations from animal cells, the rough endoplasmic reticulum (density = 1.20 g/cm^3) separates well from the Golgi vesicles (density = 1.14 g/cm^3) and from the plasma membrane (density = 1.12 g/cm^3). (The higher density of the rough endoplasmic reticulum is due largely to the ribosomes bound to it.) This method also works well for resolving lysosomes, mitochondria, and peroxisomes in the initial mixed fraction obtained by differential centrifugation.

Since each organelle has unique morphological features, the purity of organelle preparations can be assessed by examination in an electron microscope. Alternatively, organelle-specific marker molecules can be quantified. For example, the protein cytochrome *c* is present only in mitochondria, so the presence of this protein in a fraction of lysosomes would indicate its contamination by mitochondria. Similarly, catalase is present only in peroxisomes; acid phosphatase, only in lysosomes; and ribosomes, only in the rough endoplasmic reticulum or the cytosol.

ORGANELLE-SPECIFIC ANTIBODIES

Cell fractions often contain more than one type of organelle even after differential and equilibrium density-gradient centrifugation. Such fractions can be further purified by immunological techniques, using monoclonal antibodies for various organelle-specific membrane proteins. One example is the purification of a particular class of cellular vesicles whose outer surface is coated with the protein *clathrin*.

An antibody to clathrin, bound to a bacterial carrier, can selectively bind these vesicles in a crude preparation of membranes, and the whole antibody complex can then be isolated by low-speed centrifugation. A recently developed technique uses tiny metallic beads coated with specific antibodies. Organelles that bind to the antibodies, and thus are linked to the metallic beads, are recovered from the preparation by adhesion to a small magnet on the side of the test tube. All cells contain a dozen or more different types of small membrane-limited vesicles of about the same size (50–100 nm in diameter) and density. Because of their similar size and density, these vesicles are difficult to separate from one another by centrifugation techniques.

Immunological techniques are particularly useful for purifying specific classes of such vesicles. Fat and muscle cells, for instance, contain a particular glucose transporter (GLUT4) that is localized to the membrane of a specific kind of vesicle. When insulin is added to the cells, these vesicles fuse with the cell-surface membrane, a process critical to maintaining the appropriate concentration of sugar in the blood. These vesicles can be purified using an antibody that binds to a segment of the GLUT4 protein that faces the cytosol.

BIOMEMBRANES

Although all biomembranes have the same basic phospholipid bilayer structure and certain common functions, each type of cellular membrane also has certain distinctive activities determined largely by the unique set of proteins associated with that membrane. The two basic categories of membrane proteins were introduced: integral proteins, all or part of which penetrate or span the phospholipid bilayer, and peripheral proteins, which do not interact with the hydrophobic core of the bilayer. In this section, we first discuss the basic principles that govern the organization of phospholipids and integral proteins in all biological membranes and then outline the functions of the plasma membrane in prokaryotes and eukaryotes.

PHOSPHOLIPIDS

The most abundant lipid components in most membranes are phospholipids, which are amphipathic molecules (i.e., they have a hydrophilic and a hydrophobic part). In *phosphoglycerides,* a principal class of phospholipids, fatty acyl side chains are esterified to two of the three hydroxyl groups in glycerol, and the third hydroxyl group is esterified to phosphate. The phosphate group is also esterified to a hydroxyl group on another hydrophilic compound, such as choline in phosphatidylcholine. Instead of choline, alcohols such as ethanolamine, serine, and the sugar derivative inositol are linked to the phosphate in other phosphoglycerides. The negative charge on the phosphate as well as the charged groups or hydroxyl groups on the alcohol esterified to it interact strongly with water. Both of the fatty acyl side chains in a phosphoglyceride may be saturated or unsaturated, or one chain may be saturated and the other unsaturated.

Sphingomyelin, a phospholipid that lacks a glycerol backbone, is found mainly in plasma membranes. Instead of a glycerol backbone, it contains *sphingosine,* an amino alcohol with a long unsaturated hydrocarbon chain. In sphingomyelin, the terminal hydroxyl group of sphingosine is esterified to phosphocholine, so its hydrophilic head is similar to that of phosphatidylcholine. Cholesterol and its derivatives constitute another important class of membrane lipids, the steroids. The basic structure of steroids is the four-ring hydrocarbon. Cholesterol, the major steroidal constituent of animal tissues, has a hydroxyl substituent on one ring. Although cholesterol is almost entirely hydrocarbon in composition, it is amphipathic because its hydroxyl group can interact with water. Cholesterol is especially abundant in the plasma membrane of mammalian cells but is absent from most prokaryotic cells. As much as 30 to 50 percent of the lipids in plant plasma membranes consists of cholesterol and certain steroids unique to plants.

Carbohydrates are found in many membranes, covalently bound either to proteins as constituents of glycoproteins or to lipids as constituents of glycolipids. Bound carbohydrates increase the hydrophilic character of lipids and proteins and help to stabilize the conformations of many membrane proteins. The simplest glycolipid, *glucosylcerebroside,* contains a single glucose unit attached to a ceramide.

Every Cellular Membrane Forms a Closed Compartment and Has a Cytosolic and an Exoplasmic Face. Phospholipids of the composition present in cells spontaneously form symmetric sheetlike phospholipid bilayers, which are two molecules thick. The hydrocarbon side chains in each leaflet form a hydrophobic core that is 3 – 4 nm thick in most biomembranes. The various phospholipids differ in the charge carried by the polar head groups at neutral pH: some phosphoglycerides (e.g., phosphatidylcholine and phosphatidylethanolamine) have no net electric charge; others (e.g., phosphatidylglycerol and phosphatidylserine) have a net negative charge. Nonetheless, the polar head groups in all phospholipids can pack together into the characteristic bilayer structure. Sphingomyelins are similar in shape to phosphoglycerides and can form mixed bilayers with them.

Perhaps the most important lesson gleaned from the study of pure phospholipid bilayer membranes is that they spontaneously seal to form closed structures that separate two aqueous compartments. Were a phospholipid bilayer to form a sheet with ends in which the hydrophobic interior were in contact with water, it would be unstable; thus a spherical structure with no ends is the most stable state of a phospholipid bilayer.

Similarly, all cellular membranes are closed structures, surrounding the cell itself or individual compartments. Cellular membranes thus have an *internal face* (the side oriented toward the interior of the compartment) and an *external face* (the side presented to the environment). Because most organelles are surrounded by a single bilayer membrane, it is also useful to speak of the cytosolic face and exoplasmic face of the membrane, the cytosol being the part of the cytoplasm outside of organelles. Thus the exoplasmic face of such organelles faces inward. Similarly, the exoplasmic face of the plasma membrane is directed away from the cytosol, in this case toward the extracellular space, and defines the outer limit of the cell. Some organelles, such as the nucleus, mitochondrion, and chloroplast, are surrounded by two membranes; in these cases, the exoplasmic surface faces the lumen, or space, between the two membranes.

TYPES OF PHOSPHOLIPID BILAYER

A typical cell contains myriad types of membranes, each in turn bearing unique properties bestowed by its particular mix of lipids and proteins. When samples of plasma, nuclear, and mitochondrial membranes are prepared, using the cell-fractionation techniques we have already described, these preparations are often contaminated with the membranes of many other organelles. (The plasma membrane of human erythrocytes, however, can be isolated in near purity because these cells contain no internal membranes.) And all cellular membranes, regardless of their source, possess enormously varied protein-to-lipid ratios. The inner mitochondrial membrane, for example, is 76 percent protein; the myelin membrane, only 18 percent. The high phospholipid content of myelin allows it to electrically insulate the nerve cell from its environment.

Given the variable composition of cellular membranes,

how certain are we that the phospholipid bilayer structure is common to all biomembranes? One piece of evidence is that many of the physical properties of pure phospholipid bilayers are similar to those of natural cellular membranes. Another is that either a single species of phospholipid, or a mixture of phospholipids with a composition approximating that found in natural membranes, spontaneously forms either planar bilayers or liposomes when dispersed in aqueous solutions.

Perhaps the best evidence for the bilayer structure comes from low-angle x-ray diffraction analysis of the multimembrane myelin sheath, which is elaborated by Schwann cells and covers and insulates many mammalian nerve cells. The myelin sheath, which is a series of stacked membranes, is the major membrane component of such nerves and can be separated from other cellular membranes in a pure state, permitting direct physical and chemical analyses. X-ray diffraction analysis of these stacked plasma membranes has revealed a regular variation in density that is consistent with a bilayer organization of each membrane unit. In this organization, protein is located mainly on either side of the membrane, which has hydrophilic external faces and a central region of almost pure low-density hydrocarbon. Although some polypeptide segments pass through the lipid bilayer, these make up less than 10 percent of the inner mass of the membrane and are not detected in this type of analysis.

Electron microscopy of thin membrane sections stained with osmium tetroxide, which binds strongly to the polar head groups of phospholipids, provides the most direct evidence for the universality of the bilayer structure. A cross section of all single membranes stained with osmium tetroxide looks like a railroad track: two thin dark lines (the stain – head group complexes) with a uniform light space of about 2 nm (the hydrophobic tails) between them. Although some osmium tetroxide may bind to double bonds in the hydrophobic fatty acyl chains, most of it binds to the polar head groups.

INTEGRAL PROTEINS AND GLYCOLIPIDS

The spaces inside and outside the closed compartments

formed by cellular membranes usually have very different compositions. Such *asymmetry* is an essential aspect of the structure and function of biological membranes, and is reflected in the asymmetric structure of all integral membrane proteins. That is, each type of integral membrane protein has a single, specific orientation with respect to the cytosolic and exoplasmic faces of a cellular membrane, and all molecules of any particular integral membrane protein share this orientation. This absolute asymmetry in protein orientation confers different properties on the two membrane faces. Proteins have never been observed to flip-flop across a membrane; such movement, involving a transient movement of hydrophilic amino acid and sugar residues through the hydrophobic interior of the membrane, would be energetically unfavorable. Accordingly, the asymmetry of a membrane protein, which is established during its biosynthesis and insertion into a membrane, is maintained throughout the protein's lifetime.

Both glycoproteins and glycolipids are especially abundant in the plasma membrane of eukaryotic cells but are absent from the inner mitochondrial membrane, the chloroplast lamellae, and several other intracellular membranes. Almost invariably, attached carbohydrates are localized to the exoplasmic membrane face. Glycolipids are always found in the exoplasmic leaflet of membranes and are situated mainly, but not exclusively, on the surface membrane of cells. As with glycoproteins, their polar carbohydrate chains face outward toward the environment and away from the cell.

PHOSPHOLIPID COMPOSITION

Most kinds of phospholipid, as well as cholesterol, are generally present in both membrane leaflets, although they are often more abundant in one or the other. For instance, in plasma membranes from human erythrocytes and certain canine kidney cells grown in culture, almost all the sphingomyelin and phosphatidylcholine, both of which have a positively charged head group are found in the exoplasmic leaflet. In contrast, lipids with neutral or negative polar head groups (e.g., phosphatidylethanolamine, phosphatidylserine, and phosphatidylinositol) are preferentially located in the cytosolic leaflet. Phosphorylated forms of phosphatidylinositol are cleaved as a result of cell stimulation by certain hormones, generating in the cytosol soluble forms of the "head groups" that affect many aspects of cellular metabolism.

The relative abundance of a particular phospholipid in the two leaflets of a plasma membrane can be determined based on its susceptibility to hydrolysis by *phospholipases,* enzymes that cleave the phosphoester bonds that connect the phospholipid head groups. Phospholipids in the cytosolic leaflet are resistant to hydrolysis by phospholipases added to the external medium, because the enzymes cannot penetrate to the cytosolic face of the plasma membrane. It is not clear how these differences in lipid composition of the two leaflets arise. One possibility is that certain lipids bind to specific protein domains that occur preferentially in one membrane leaflet.

MOST LIPIDS AND INTEGRAL PROTEINS

In both pure phospholipid bilayers and natural membranes, thermal motion permits phospholipid and glycolipid molecules to rotate freely around their long axes and to diffuse laterally within the membrane leaflet. Because such movements are lateral or rotational, the fatty acyl chains remain in the hydrophobic interior of the membrane. In both natural and artificial membranes, a typical lipid molecule exchanges places with its neighbors in a leaflet about 10^7 times per second and diffuses several micrometers per second at 37 °C. At this rate, a lipid could diffuse the length of a typical bacterial cell (»1 mm) in only 1 second and the length of an animal cell in about 20 seconds.

In pure phospholipid bilayers, phospholipids do not migrate, or flip-flop, from one leaflet of the membrane to the other. In some natural membranes, however, they occasionally do so, catalyzed by certain membrane proteins called *flippases*). Energetically, such movements are extremely unfavorable, because the polar head of a phospholipid must be transported through the hydrophobic interior of the membrane. Various experiments have shown that many integral membrane proteins, like phospholipids, float quite freely within the plane of a natural membrane.

In one such study two different cells (e.g., mouse and human fibroblasts) are fused and the movement of their distinct surface proteins is then monitored at various times after incubation at 37 °C. Such experiments suggest that many integral proteins are free to diffuse in a sea of lipid in the two-dimensional space of the membrane. According to this concept, known as the *fluid mosaic model,* the membrane is viewed as a two-dimensional mosaic of laterally mobile phospholipid and protein molecules. As discussed some integral membrane proteins consist of two or more noncovalently linked subunits; such multimeric membrane proteins float as a unit in the lipid.

The lateral movements of surface proteins and lipids can be quantified by a technique called *fluorescence recovery after photobleaching* (FRAP). With this method the rate at which surface protein or lipid molecules move — the diffusion coefficient — can be determined, as well as the proportion of the molecules that are laterally mobile. FRAP studies with fluorescent-labeled phospholipids have shown that in fibroblast plasma membranes, all the phospholipids are freely mobile over distances of about 0.5 µm, but most cannot diffuse over much longer distances.

These findings suggest that protein-rich regions of the plasma membrane, about 1 µm in diameter, separate lipid-rich regions containing the bulk of the membrane phospholipid. Phospholipids are free to diffuse within such a region but not from one lipid-rich region to an adjacent one. Furthermore, the rate of lateral diffusion of lipids in the plasma membrane is nearly an order of magnitude slower than in pure phospholipid bilayers: diffusion constants of 10^{-8} cm^2/s and 10^{-7} cm^2/s are characteristic of the plasma membrane and a lipid bilayer, respectively. This difference suggests that lipids may be tightly but not irreversibly bound to certain integral proteins in some membranes.

Numerous experiments similar to those just discussed

have shown that depending on the cell type, 30 – 90 percent of all integral proteins in the plasma membrane are freely mobile. Immobile proteins are permanently attached to the underlying cytoskeleton. The lateral diffusion rate of a mobile protein in an intact membrane is generally 10 – 30 times lower than that of the same protein embedded in synthetic liposomes. These findings suggest that the mobility of integral proteins in intact membranes is restricted by interactions with the rigid submembrane cytoskeleton. Clearly, such interactions would have to be broken and remade as mobile proteins diffuse laterally in the plasma membrane.

FLUIDITY OF MEMBRANES

One consequence of the packing of the fatty acyl chains within the centre of a phospholipid bilayer is an abrupt change in its physical properties over a very narrow temperature range. For example, when a suspension of liposomes or a planar bilayer composed of a single type of phospholipid is heated, it passes from a highly ordered, gel-like state to a more mobile fluid state. During this *phase transition,* a relatively large amount of heat (thermal energy) is absorbed over a narrow temperature range; the midpoint of this range is the "melting temperature" of the bilayer.

In general, lipids with short or unsaturated fatty acyl chains undergo the phase transition at lower temperatures than do lipids with long or saturated chains. Compared with long chains, short chains have less surface area to form van der Waals interactions with one another. Since the gel state is stabilized by these interactions, short-chain lipids melt at lower temperatures than long-chain lipids. Likewise, the kinks in unsaturated fatty acyl chains result in their forming less stable van der Waals interactions with other lipids than do saturated chains. As a result, unsaturated lipids maintain a more random, fluid state at lower temperatures than lipids with saturated fatty acyl chains.

The hydrophobic interior of natural membranes generally has a low viscosity and a fluidlike, rather than gel-like, consistency. Maintenance of this bilayer fluidity appears to be essential for normal cell growth and reproduction. All cell membranes contain a mixture of different fatty acyl chains and are fluid at the temperature at which the cell is grown. Animal and bacterial cells adapt to a decrease in growth temperature by increasing the proportion of unsaturated to saturated fatty acids in the membrane, which tends to maintain a fluid bilayer at the reduced temperature.

Membrane cholesterol is another major determinant of bilayer fluidity. Cholesterol is too hydrophobic to form a sheet structure on its own, but it is intercalated (inserted) among phospholipids. Its polar hydroxyl group is in contact with the aqueous solution near the polar head groups of the phospholipids; the steroid ring interacts with and tends to immobilize their fatty acyl chains. The net effect of cholesterol on membrane fluidity varies, depending on the lipid composition. Cholesterol restricts the random movement of the polar heads of the fatty acyl chains, which are closest to the outer surfaces of the leaflets, but it separates and disperses their tails, causing the inner regions of the bilayer to become slightly more fluid. At the high concentrations found in eukaryotic plasma membranes, cholesterol tends to make the membrane less fluid at growth temperatures near 37 °C. Below the temperature that causes a phase transition, cholesterol keeps the membrane in a fluid state by preventing the hydrocarbon fatty acyl chains of the membrane lipids from binding to one another, thereby offsetting the drastic reduction in fluidity that would otherwise occur at low temperatures.

MEMBRANE LEAFLETS

When a frozen tissue specimen is fractured by a sharp blow, the fracture line frequently runs through the hydrophobic interior of cell membranes, separating the two phospholipid leaflets. Integral membrane proteins generally remain associated with one or the other leaflet of membranes subjected to this *freeze-fracturing* technique. The fractured specimen then is placed in a vacuum and the surface ice is removed by sublimation, a technique called *deep etching* or *freeze etching*. After metal shadowing with platinum and carbon, the organic material is removed by acid, leaving a carbon-metal replica of the membrane leaflet.

Electron microscopy of membrane samples prepared by these techniques reveals numerous protuberances, most of which are membrane proteins. In deep-etching studies, the cytoplasmic face of a membrane is customarily called the *P (protoplasmic) face* and the exoplasmic face is the *E face*. It is not unusual for most or all protuberances to be on one of the two surfaces and their mirror images, in the form of pits or holes, to be on the other. This may occur because the integral proteins are bound more tightly to the lipids in one leaflet than to those in the other.

THE PLASMA MEMBRANE

In all cells, the plasma membrane has several essential functions. These include transporting nutrients into and metabolic wastes out of the cell; preventing unwanted materials in the extracellular milieu from entering the cell; preventing loss of needed metabolites and maintaining the proper ionic composition, pH (7.2), and osmotic pressure of the cytosol. To carry out these functions, the plasma membrane contains specific *transport proteins* that permit the passage of certain small molecules but not others. Several of these proteins use the energy released by ATP hydrolysis to pump ions and other molecules into or out of the cell against their concentration gradients. Small charged molecules such as ATP and amino acids can diffuse freely within the cytosol but are restricted in their ability to leave or enter it across the plasma membrane.

In addition to these universal functions, the plasma membrane has other crucial roles in multicellular organisms. Few of the cells in multicellular plants and animals exist as isolated entities; rather, groups of cells with related specializations combine to form tissues. Specialized areas of the plasma membrane contain proteins and glycolipids that form specific contacts and junctions between cells to strengthen tissues and to allow the exchange of metabolites between cells. Other proteins in the plasma membrane act as anchoring points for many of the cytoskeletal fibers that permeate the cytosol, imparting shape and strength to cells.

Surrounding most animal cells is a mixture of fibrous proteins and polysaccharides collectively called the extracellular matrix. This viscous, water-filled matrix provides a bedding on which most sheets of epithelial cells or small glands lie. Proteins in the plasma membrane anchor cells to many of the matrix components, adding to the strength and rigidity of many tissues. In addition, enzymes bound to the plasma membrane catalyze reactions that would occur with difficulty in an aqueous environment. The plasma membrane of many types of eukaryotic cells also contains receptor proteins that bind specific signaling molecules (e.g., hormones, growth factors, neurotransmitters), leading to various cellular responses. These membrane proteins, which are critical for cell development and functioning, are described in later chapters.

Unlike animal cells, plant cells are surrounded by a cell wall and lack the extracellular matrix found in animal tissues. As a plant cell matures, new layers of wall are laid down just outside the plasma membrane, which is intimately involved in the assembly of cell walls. The walls are built primarily of cellulose, a rodlike polysaccharide formed from b(1→ 14)-linked glucose monomers. The cellulose molecules aggregate, by hydrogen bonding, into bundles of fibers; other polysaccharides within the wall cross-link the cellulose fibers. In woody plants, a complex water-insoluble polymer of phenol and other aromatic monomers, called *lignin,* imparts strength and rigidity to the cell walls. Other chemicals also are found in the walls of various plant cells; for example, waxes prevent plant tissues and proteins from drying out.

Like the entire cell, each organelle in eukaryotic cells is bounded by a membrane containing a unique set of proteins essential for its proper functioning. In the next section, we discuss the structure and function of the main organelles found in eukaryotic cells.

ORGANELLES OF THE EUKARYOTIC CELL

The various techniques described earlier have led to an

appreciation of the highly organized internal structure of eukaryotic cells, marked by the presence of many different organelles. Here we present a brief overview of the major organelles. Unique proteins in the interior and membranes of each type of organelle largely determine its specific functional characteristics. Later chapters will examine the key roles that different organelles and the cytosol play in the functioning of eukaryotic cells.

ACIDIC ORGANELLES

Lysosomes provide an excellent example of the ability of intracellular membranes to form closed compartments in which the composition of the *lumen* (the aqueous interior of the compartment) differs substantially from that of the surrounding cytosol. Found in animal cells, lysosomes are bounded by a single membrane and are responsible for degrading certain components that have become obsolete for the cell or organism. In some cases, materials taken into a cell by endocytosis or phagocytosis also are degraded in lysosomes. Endocytosis refers to the process by which extracellular materials are taken up by invagination of a segment of the plasma membrane to form a small membrane-bounded vesicle (endosome). In phagocytosis, relatively large particles are enveloped by the plasma membrane and internalized.

Lysosomes contain a group of enzymes that degrade polymers into their monomeric subunits. For example, nucleases degrade RNA and DNA into their mononucleotide building blocks; proteases degrade a variety of proteins and peptides; phosphatases remove phosphate groups from mononucleotides, phospholipids, and other compounds; still other enzymes degrade complex polysaccharides and lipids into smaller units. *Tay-Sachs disease* is caused by a defect in one enzyme catalyzing a step in the lysosomal breakdown of certain glycolipids called gangliosides, which are abundant in nerve cells—with devastating consequences. The symptoms of this inherited disease usually are evident before the age of 1. Affected children commonly become demented and blind by age 2, and die before their third birthday. Nerve cells from such children are greatly enlarged with swollen lipid-filled lysosomes.

All the lysosomal enzymes work most efficiently at acid pH values and collectively are termed *acid hydrolases*. A hydrogen ion pump and a Cl^- channel protein in the lysosomal membrane maintain the pH of the interior at ≈4.8. The pump hydrolyzes ATP and uses the released free energy to pump H^+ ions from the cytosol into the lumen of the lysosome; the Cl^- channel allows Xλ ions to enter. Together they transport HCl. The acid pH helps to denature proteins, making them accessible to the action of the lysosomal hydrolases, which themselves are resistant to acid denaturation. Lysosomal enzymes are poorly active at the neutral pH values of cells and most extracellular fluids. Thus if a lysosome releases its contents into the cytosol, where the pH is between 7.0 and 7.3, little degradation of cytosolic components takes place.

Lysosomes vary in size and shape, and several hundred may be present in a typical animal cell. In effect, they function as sites where various materials to be degraded collect. *Primary lysosomes* are roughly spherical and do not contain obvious particulate or membrane debris. *Secondary lysosomes,* which are larger and irregularly shaped, appear to result from the fusion of primary lysosomes with other membrane organelles; they contain particles or membranes in the process of being digested. The process by which an aged organelle is degraded in a lysosome is called *autophagy* ("eating oneself").

PLANT VACUOLES

Most plant cells contain at least one membrane-limited internal vacuole. The number and size of vacuoles depend on both the type of cell and its stage of development; a single vacuole may occupy as much as 80 percent of a mature plant cell. Plant cells store water, ions, and nutrients such as sucrose and amino acids within these vacuoles. We will see how such materials are accumulated in vacuoles. Vacuoles also act as receptacles for waste products and excess salts taken up by the plant and may function similarly to lysosomes in animal cells. Like lysosomes, vacuoles have an acidic pH, maintained by a proton pump and a Cl^- channel protein in the vacuole membrane, and contain a battery of degradative enzymes. Similar storage vacuoles are found in green algae and many microorganisms such as yeast.

Like most cellular membranes, the vacuolar membrane is permeable to water but is poorly permeable to the small molecules stored within it. Because the solute concentration is much higher in the vacuole lumen than in the cytosol or extracellular fluids, water tends to move by osmotic flow into vacuoles, just as it moves into cells placed in a hypotonic medium.

This influx of water causes both the vacuole to expand and water to move into the cell from the wall, creating hydrostatic pressure, or *turgor,* inside the cell. This pressure is balanced by the mechanical resistance of the cellulose-containing cell wall that surrounds plant cells. Most plant cells have a turgor of 5 – 20 atmospheres (atm); their cell walls must be strong enough to react to this pressure in a controlled way. Unlike animal cells, plant cells can elongate extremely rapidly—at rates of 20–75 μm/h. This elongation, which usually accompanies plant growth, occurs when a segment of the somewhat elastic cell wall stretches under the pressure created by water taken into the vacuole.

PEROXISOMES DEGRADE FATTY ACIDS AND TOXIC COMPOUNDS

All animal cells (except erythrocytes) and many plant cells contain peroxisomes, a class of small organelles (≈0.2 – 1 μm in diameter) bounded by a single membrane. (*Glyoxisomes* are similar organelles found in plant seeds that oxidize stored lipids as a source of carbon and energy for growth. They contain many of the same types of enzymes as peroxisomes as well as additional ones used to convert fatty acids to glucose precursors.) Peroxisomes contain several *oxidases* — enzymes that use molecular oxygen to oxidize organic substances, in the process forming hydrogen peroxide (H_2O_2), a corrosive substance. Peroxisomes also contain copious amounts of the enzyme *catalase,* which degrades hydrogen peroxide to yield water and oxygen:

$$2H_2O_2 \xrightarrow{catalase} 2H_2O + O_2$$

In contrast to oxidation of fatty acids in mitochondria, which produces CO_2 and is coupled to generation of ATP, peroxisomal oxidation of fatty acids yields acetyl groups and is not linked to ATP formation. The energy released during peroxisomal oxidation is converted to heat, and the acetyl groups are transported into the cytosol, where they are used in the synthesis of cholesterol and other metabolites. In most eukaryotic cells, the peroxisome is the principal organelle in which fatty acids are oxidized, thereby generating precursors for important biosynthetic pathways. Particularly in liver and kidney cells, various toxic molecules that enter the bloodstream also are degraded in peroxisomes, producing harmless products.

In the human genetic disease *X-linked adrenoleukodystrophy* (ADL), peroxisomal oxidation of very long chain fatty acids is defective. The *ADL* gene encodes the peroxisomal membrane protein that transports into peroxisomes an enzyme required for oxidation of these fatty acids. Individuals with the severe form of ADL are unaffected until mid-childhood, when severe neurological disorders appear, followed by death within a few years.

Mitochondria Are the Principal Sites

Most eukaryotic cells contain many mitochondria, which occupy up to 25 percent of the volume of the cytoplasm. These complex organelles, the main sites of ATP production during aerobic metabolism, are among the largest organelles, generally exceeded in size only by the nucleus, vacuoles, and chloroplasts.

Mitochondria contain two very different membranes, an outer one and an inner one, separated by the intermembrane space. The outer membrane, composed of about half lipid and half protein, contains proteins that render the membrane permeable to molecules having molecular weights as high as 10,000. In this respect, the outer membrane is similar to the outer membrane of gram-negative bacteria. The inner membrane, which is much less permeable, is about 20 percent lipid and 80 percent protein — a higher proportion of protein than occurs in other cellular membranes. The surface area of the inner membrane is greatly increased by a large number of infoldings, or *cristae,* that protrude into the *matrix,* or central space.

In nonphotosynthetic cells, the principal fuels for ATP synthesis are fatty acids and glucose. The complete aerobic degradation of glucose to CO_2 and H_2O is coupled to synthesis of as many as 36 molecules of ATP. In eukaryotic cells, the initial stages of glucose degradation occur in the cytosol, where two ATP molecules per glucose molecule are generated. The terminal stages, including those involving phosphorylation coupled to final oxidation by oxygen, are carried out by enzymes in the mitochondrial matrix and cristae.

As many as 34 ATP molecules per glucose molecule are generated in mitochondria, although this value can vary because much of the energy released in mitochondrial oxidation can be used for other purposes (e.g., heat generation and the transport of molecules into or out of the mitochondrion), making less energy available for ATP synthesis. Similarly, virtually all the ATP formed during the oxidation of fatty acids to CO_2 is generated in the mitochondrion. Thus the mitochondrion can be regarded as the "power plant" of the cell.

CHLOROPLASTS

Except for vacuoles, chloroplasts are the largest and most characteristic organelles in the cells of plants and green algae. They can be as long as 10 μm and are typically 0.5 – 2 μm thick, but they vary in size and shape in different cells, especially among the algae.

Like the mitochondrion, the chloroplast is surrounded by an outer and an inner membrane. Chloroplasts also contain an extensive internal system of interconnected membrane-limited sacs called thylakoids, which are flattened to form disks; these often are grouped in stacks called *grana* and embedded in a matrix, the *stroma*. The thylakoid membranes contain green pigments (chlorophylls) and other pigments and enzymes that absorb light and generate ATP during photosynthesis. Part of this ATP is used by enzymes located in the stroma to convert CO_2 into three-carbon intermediates; these are then exported to the cytosol and converted to sugars.

Perhaps surprisingly, the molecular mechanisms by which ATP is formed in mitochondria and chloroplasts are very similar. Chloroplasts and mitochondria share other features: Both often migrate from place to place within cells and also contain their own DNA, which encodes some of the key organellar proteins. The proteins encoded by mitochondrial or chloroplast DNA are synthesized on ribosomes within the organelles. However, most of the proteins in each organelle are encoded in nuclear DNA and are synthesized in the cytosol; these proteins then are incorporated into the organelles.

THE ENDOPLASMIC RETICULUM

Generally, the largest membrane in a eukaryotic cell encloses the endoplasmic reticulum (ER) — a compartment comprising a network of interconnected, closed, membrane-bounded vesicles. The endoplasmic reticulum has a number of functions in the cell but is particularly important in the synthesis of many membrane lipids and proteins. The *smooth endoplasmic reticulum* is smooth because it lacks ribosomes; regions of the *rough endoplasmic reticulum* are studded with ribosomes.

THE SMOOTH ENDOPLASMIC RETICULUM

The synthesis of fatty acids and phospholipids occurs in the smooth ER. Although many cells have very little smooth ER, this organelle is abundant in hepatocytes. Enzymes in the smooth ER of the liver modify or detoxify hydrophobic chemicals such as pesticides and carcinogens by chemically converting them into more water-soluble, conjugated products that can be secreted from the body. High doses of such compounds result in a large proliferation of the smooth ER in liver cells.

THE ROUGH ENDOPLASMIC RETICULUM

Ribosomes bound to the rough ER synthesize certain membrane and organelle proteins and virtually all proteins to be secreted from the cell. The ribosomes that fabricate secretory proteins are bound to the rough ER by the nascent polypeptide chain of the protein. As the growing secretory polypeptide emerges from the ribosome, it passes through the rough ER membrane, with the help of specific proteins in the membrane. The newly made secretory proteins accumulate in the lumen (inner cavity) of the rough ER before being transported to their next destination.

All eukaryotic cells contain a discernible amount of rough ER because it is needed for the synthesis of plasma-membrane proteins and proteins of the extracellular matrix. Rough ER is particularly abundant in cells that are specialized to produce secreted proteins. For example, plasma cells produce antibodies, which circulate in the bloodstream, and pancreatic acinar cells synthesize digestive enzymes, which are transported to the intestine via a series of progressively larger ducts. In both types of cells, a large part of the cytosol is filled with rough ER.

GOLGI VESICLES

Several minutes after proteins are synthesized in the rough ER, most of them leave the organelle within small membrane-bounded transport vesicles. These vesicles, which bud off from regions of the rough ER not coated with ribosomes, carry the proteins to the luminal cavity of another membrane-limited organelle, the Golgi complex, a series of flattened sacs located near the nucleus in many cells.

Three-dimensional reconstructions from serial sections of a Golgi complex reveal a series of flattened membrane vesicles or sacs, surrounded by a number of more or less spherical membrane vesicles. The stack of flattened Golgi sacs has three defined regions — the *cis,* the *medial,* and the *trans*. Transfer vesicles from the rough ER fuse with the cis region of the Golgi complex, where they deposit their proteins. These proteins then progress from the cis to the medial to the trans region.

Within each region are different enzymes that modify secretory and membrane proteins differently, depending on their structures and their final destinations.

After secretory proteins are modified in the Golgi sacs, they are transported out of the complex by a second set of transport vesicles, which seem to bud off the trans side of the Golgi complex. Some of these transport vesicles, termed *coated vesicles,* are surrounded by an outer protein cage composed primarily of the fibrous protein clathrin. Some vesicles contain membrane proteins destined for the plasma membrane; others, proteins for lysosomes or for other organelles. How intracellular transport vesicles "know" which membranes to fuse with and where to deliver their contents is also discussed.

THE DOUBLE-MEMBRANED NUCLEUS

The nucleus, the largest organelle in eukaryotic cells, is surrounded by two membranes, each one a phospholipid bilayer containing many different types of proteins. The inner nuclear membrane defines the nucleus itself. In many cells, the outer nuclear membrane is continuous with the rough endoplasmic reticulum, and the space between the inner and outer nuclear membranes is continuous with the lumen of the rough endoplasmic reticulum.

The two nuclear membranes appear to fuse at the nuclear pores. The distribution of nuclear pores is particularly vivid when the nucleus is viewed by the freeze-fracture technique described earlier. Constructed of a specific set of membrane proteins, these ringlike pores function as channels that regulate the movement of material between the nucleus and the cytosol.

In a growing or differentiating cell, the nucleus is metabolically active, producing DNA and RNA. The latter is exported through nuclear pores to the cytoplasm for use in protein synthesis. In mature erythrocytes from nonmammalian vertebrates and other types of "resting" cells, the nucleus is inactive or dormant and minimal synthesis of DNA and RNA takes place.

In a nucleus that is not dividing, the chromosomes are dispersed and not thick enough to be observed in the light microscope. Only during cell division are chromosomes visible by light microscopy. However, a suborganelle of the nucleus, the nucleolus, is easily recognized under the light microscope. Most of the cell's ribosomal RNA is synthesized in the nucleolus; some ribosomal proteins are added to ribosomal RNAs within the nucleolus as well. The finished or partly finished ribosomal subunit passes through a nuclear pore into the cytosol.

In the electron microscope, the nonnucleolar regions of the nucleus, called the *nucleoplasm,* can be seen to have areas of high DNA concentration, often closely associated with the nuclear membrane. Fibrous proteins called lamins form a two-dimensional network along the inner surface of the inner membrane, giving it shape and apparently binding DNA to it. The breakdown of this network occurs early in cell division.

THE CYTOSOL

Because sections for standard electron microscopy must be thinner than 0.1 mm, fibers in the cytosol, which may be several microns in length, appear as long elements only in sections that by chance happen to be in the plane of the fibre bundles. Serial sectioning of a tissue sample can compensate for these shortcomings by tracing a fibre from one image into the next to reconstruct its threedimensional architecture. Because 200 sections are needed to examine a cell 20 mm thick, serial sectioning is a tedious technique. However, sections up to 1 mm thick can be viewed in high-voltage electron microscopes, considerably reducing the number of sections needed to reconstruct three-dimensional images.

Transmission electron micrographs of cytosolic fibers obtained from unsectioned cells reveal an extensive network of microfilaments, microtubules, and intermediate filaments. These cytoskeletal fibers crisscross one another in complex patterns so that different types of fibers contact one another at many points. In cultured cells, actin microfilaments often occur in bundles of long fibers that appear to be connected by small fibrous proteins.

The cytosol of many cells also contains *inclusion bodies,*

granules that are not bounded by a membrane. For instance, muscle cells and hepatocytes contain cytosolic granules of glycogen a glucose polymer that functions as a storage form of usable cellular energy. In well-fed animals, glycogen can account for as much as 10 percent of the wet weight of the liver. The cytosol of the specialized fat cells in adipose tissue contains large droplets of almost pure triacylglycerols, a storage form of fatty acids.

In addition, the cytosol is a major site of cellular metabolism and contains a large number of different enzymes. About 20 – 30 percent of the cytosol is protein, and from a quarter to half of the total protein within cells is in the cytosol. Because of the high concentration of cytosolic proteins, organized complexes of proteins can form even if the energy that stabilizes them is weak. Many investigators believe that the cytosol is highly organized, with most proteins either bound to fibers or otherwise localized in specific regions.

Chapter 11

Fusion of Two Techniques

THE NEXT BIG LEAP

The next big leap in medical science depends largely on how fast and how well scientists succeed in understanding the activity of thousands of tiny organs in a new world which they have only recently begun to explore -the world of the living cell.

This miniature world holds the key to the major health problems of today: cancer, atherosclerosis, genetic diseases, diabetes, mental illness. It is no longer a matter of fighting bacteria or viruses, now that vaccines and antibiotics can prevent or cure so many infectious diseases. The illnesses that still defeat us are far more insidious, for they result from disorders within the human cell, the basic unit of the body. To control these, we need to know much more about what goes on inside the cell.

In the past 30 years, researchers have created a new field, modern cell biology. Through an ingenious combination of electron microscopy, biochemical separations and analyses of cell parts, X-ray crystallography and other means, they have revealed the cell as enormously complex and tightly organized -- a strange, watery world in which diaphanous particles of various sizes and shapes float about, in rapid motion, engaged in thousands of chemical reactions. Each cell has its own power plants; its own digestive system; its own factories for making proteins and other essential molecules. Most important, it has an intricate communications network through which it can regulate its own activities (for example, "decide" when to reproduce) as well as receive messages from other cells, sense changes in its environment and send out messages of its own.

The discovery of this teeming world within the cell and the study of its organization are leading scientists to develop some entirely new concepts of health and disease. Cancer, for instance, can now be viewed -- and studied -- as a derangement of the cell cycle. No matter what organ of the body is attacked by cancer, the disease always involves tissue whose cells have lost the ability to sense when they should stop multiplying. Normal cells stop reproducing upon contact with other cells, in a process called "contact inhibition." But cancer cells go right on growing and multiplying, even piling up in layers that become tumors, as if their surface membranes could no longer perceive signals from nearby cells. Instead of staying in the part of the body for which they were specialized, they often slide away and migrate to other parts of the body, to form new tumors. The control of cancer may well depend on unraveling these mechanisms.

Other incurable or chronic diseases may some day be related to defects of the cell's mitochondria, the curious, semi-independent little organs (organelles) in the cell which act as the cell's power plants.

Mitochondria reproduce themselves with their own genetic material and many scientists believe that they are descended from bacteria-like parasites which infected a primitive cell billions of years ago and then stayed on, in a useful symbiotic relationship with other parts of the cell. Without the energy produced by the mitochondria, cells could neither move nor synthesize the chemicals they need nor, in the case of muscle cells, fuel their contractions. The study of mitochondria may thus provide us with new ways of preventing or treating a wide variety of diseases -- for example, certain diseases involving muscle cells.

Researchers are also beginning to zero in on the genetic factors which often play such an important role in disease. Many disorders of organelles are hereditary, since the information for making these organelles comes from the genes, the units of hereditary material which are found in the nuclei of cells. Some genetic diseases have been traced to defects in a single gene. In many more diseases, however, including such widespread ones as diabetes, coronary heart disease and schizophrenia, several genes as well as various environmental factors appear to be involved.

Despite all the new clues and the magnificent achievements of the last 30 years, scientists are only beginning to approach the stage where their research will have a major impact on human health. The cell used to be almost virgin territory. The pioneers of cell biology have succeeded in mapping out and analyzing its major features. But they are still very far from understanding the mechanisms which control the organelles' varied activities. They do not know, for example, what regulates the series of orderly changes involved in normal cell growth and development.

It is one of the greatest mysteries in biology that although each human being starts life as a single cell and each cell then divides into two identical cells, which divide again, somehow these cells differentiate in a highly ordered fashion so that some make an eye, some make hair, some make bone or blood and others make nerves or skin tissue. How do these cells know what to do? Why do certain kinds of cells die, never to be replaced, while others go on reproducing? What happens to our cells -- and to ourselves -- as we age?

Vast amounts of information about the cell had to be accumulated before scientists could even attempt to study such questions. It is only now that such research on cell regulation is becoming possible.

This pamphlet will describe some of the discoveries that have brought us to this breath-taking stage and then focus on one of the most exciting areas of current cell research: The study of the cell's ultra-thin but extraordinarily active surface membrane, which plays an important role in many aspects of health and disease. Much of this work is being supported by the Cellular and Molecular Basis of Disease Programme of the National Institute of General Medical Sciences (NIGMS) which is part of the National Institutes of Health (NIH), Bethesda, Maryland.

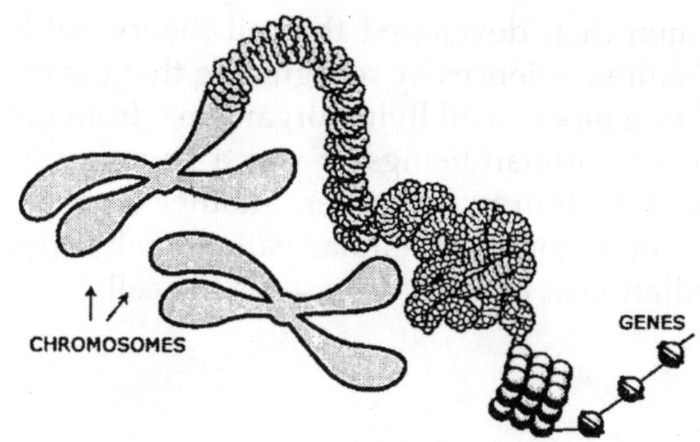

Fig. The Variety Of Human Cells

The single cell with which we start life divides again and again into many kinds of specialized cells whose structure varies according to their function. Some nerve cell fibers extend over 3 feet in length to reach from spine to toe. The orderly structure of muscle cells is shown here in such detail that if the entire muscle cell were drawn on the same scale as this fragment, it might be 1,000 feet long.

WHAT ARE CELLS?

The idea that every human and animal body is made up of cells emerged with full force from an encounter between two German scientists in 1838. Nearly two centuries earlier, in 1665, the English physicist Robert Hooke had peered at a sliver of cork through a primitive microscope and noticed some "pores" or "cells" in it. But since these looked like cavities (they were actually the cell walls of dead cork tissue), they were believed to serve as containers for the "noble juices" or "fibrous threads" of plant life. Besides, such cells seemed to exist only in plant material.

During a dinner conversation in 1838, however, the botanist Matthias Schleiden, who had been studying plant cells and the zoologist Theodor Schwann, who had been examining the nervous tissue of animals, suddenly realized that the similarities between the structures they had been investigating were too strong to be accidental. Eventually Schwann showed that all animal tissue is composed of cells, including bone, blood, skin, muscle and glands. Even sperm and eggs are cells.

The two men then developed the cell theory, which united plant and animal sciences by recognizing that the cell was the basic building block of all living organisms, from orchids and earthworms to human beings. It was a stunning intellectual achievement. Within two decades, another German scientist, Rudolf Virchow, founded cellular pathology by showing that the immediate seat of most diseases is the cell.

Chapter 12

The Birth of Modern Cell Biology

When a Belgian physician andreas Vesalius, published the first detailed descriptions and drawings of the organs of the body in 1543, he started a Renaissance in medical science. Strictly forbidden in medieval times, the dissection of human cadavers had not been practiced by Western physicians for over a thousand years, until the 14th century. Vesalius' precise descriptions of human organs suddenly raised important questions about their functions and in attempting to answer these questions, anatomy became a scientific discipline.

Another Renaissance in medical science began roughly 30 years ago, when the electron microscope revealed that each cell has various organs of its own. These organs, or organelles, are as essential to the life of the cell as the heart, liver and brain are essential to the life of the human body. And once again, the awareness of so many different structures, each with a characteristic size, shape and function, is raising questions which challenge older concepts of health and disease.

The early 1950's saw an avalanche of discoveries about the world within the cells of animals and plants. A whole new vocabulary had to be developed for such unfamiliar structures as lysosomes, mitochondria, ribosomes, the rough endoplasmic reticulum, the smooth endoplasmic reticulum, microfilaments, microtubules, the plasma membrane and others. At first these sounded exotic and rather forbidding, but by now each organelle has emerged with a personality of its own and scientists speak of them not only with assurance but even, at times, with affection or humour. "I regret somewhat that I cannot

Despite their extraordinary variety, the cells of all species are made of the same fundamental materials: nucleic acids, proteins, lipids, cargohydrates, water and salt.

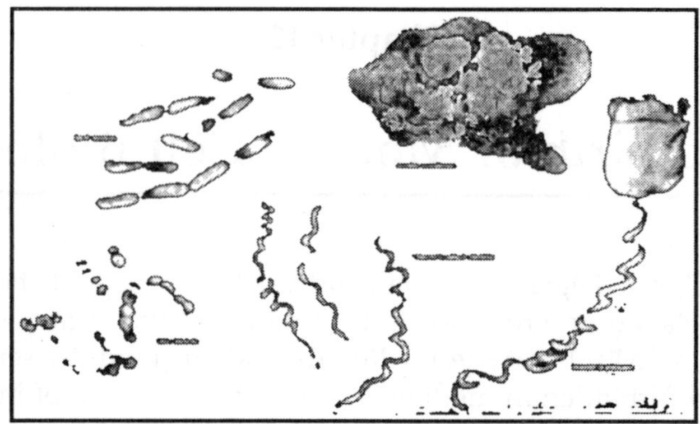

be in closer touch with my mitochondria," writes Lewis Thomas, president of the Memorial Sloan-Kettering Cancer Centre in New York, in his book, The Lives of a Cell. "If I concentrate, I can imagine that I feel them; they do not quite squirm, but there is, from time to time, a kind of tingle."

In their rush to study the new particles' anatomical organization, chemical functions and molecular structure, scientists have had to merge many disparate skills. Alone, each group of specialists could uncover only a small piece of the jigsaw puzzle, but by combining their techniques they have been able to delve further and further into the mysteries of the cell.

For example, they have begun to decipher the composition and three-dimensional structure of enzymes and other essential proteins. Enzymes are extraordinarily efficient catalysts which can speed up chemical reactions up to 100,000 times. They do this at body temperature and in a neutral solution, when chemists would have to use very high temperatures and potent chemicals, such as strong acids or bases, to achieve the same speed. We depend on enzymes to break down the food we eat and provide us with energy; to get rid of unwanted chemicals in our cells; to synthesize other proteins, nucleic acids, fats and carbohydrates; to maintain the chemical balance within our cells; and to perform hundreds of other vital functions.

While the biochemistry of cells is extremely complex, each basic step consists of a single chemical reaction, catalyzed by a specific enzyme. A single bacterium may have several thousand different enzymes and a mammal many more. Yet each of these enzyme molecules is so intricate that two decades ago it seemed beyond the powers of chemistry and physics to understand its structure. Although the structures of only a small number of enzymes have been uncovered to date, this achievement -- which depended on X-ray crystallography and other new methods -- has been one of the triumphs of molecular biology. It is changing our view of key operations within the cell, such as the energy transformations within the mitochondrion.

The joint effort of electron microscopists, biochemists, physical chemists, X-ray crystallographers, physiologists and geneticists, brought to bear on such problems, has produced modern cell biology.

GENERAL CHARACTERISTICS OF THE CELL

Each of us starts life as a single cell, a microscopic package which contains directions for everything that we can become.

This cell is defined and separated from the rest of the world by a membrane, a transparent film so thin that it could not be seen under any light microscope. For years scientists debated whether such a membrane truly existed -- perhaps it was just an illusion? Despite its ethereal appearance, however, the surface membrane is not only real but exceedingly powerful, controlling everything that goes in and out of the cell and relaying vital messages. Similar membranes enclose or make up a large number of the cell's organelles.

"It takes a membrane to make sense out of disorder in biology," writes Lewis Thomas. "You have to be able to catch energy and hold it, storing precisely the needed amount and releasing it in measured shares. A cell does this and so do the organelles inside it.... you can only transact this business with membranes in our kind of world."

Sheltered by the cell's surface membrane (which is also called the plasma membrane), different configurations of organelles move about in a watery environment. The number and kinds of organelles in each cell depend on its function.

Now that researchers have become acquainted with these organelles, they can often guess at a cell's specialty from its structure: A cell that contains many mitochondria, for instance, must be involved in producing large quantities of energy, (It was once calculated that cells deepens as the cells move apart.

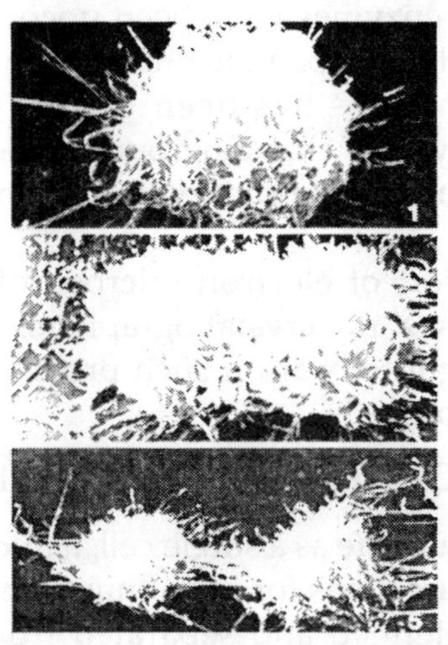

A human cell Splits: These SEM (scanning electron microscope) photographs show one cell dividing into two identical daughter cells. The cleavage furrow between the daughter. The hair-like projections on the cell surface are microvilli, cytoplasmic extensions which seem to be involved in cell motion.

Bumble-bees cannot fly, since there seemed no way for their short wings to get enough lift -- yet obviously they do fly, despite their weight and one reason is that the cells in their wing muscles contain exceptionally large numbers of mitochondria which supply them with the energy to beat at enormous speed. A cell in which ribosomes are prominent must be synthesizing proteins. A cell with a great many lysosomes must be destroying wastes. And so on.

The single cell with which we start life divides again and again. As Daniel Mazia, Professor of Zoology at the University of California, puts it, the story of the cell cycle is "Double or nothing. With few exceptions, a living cell either reproduces or dies; the principle is so simple that no one has bothered to call it a principle. A cell is born in the division of a parent cell. It then doubles in every respect: in every part, in every kind of molecule, even in the amount of water it contains."

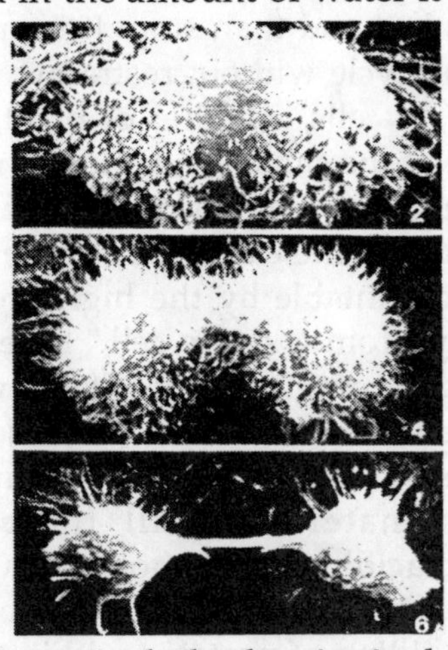

The process is particularly clear in single-celled organisms such as amoebae, which may live forever through doublings. Each cell annihilates its individual existence in the production of two daughter cells, which repeat the process roughly 40 hours later, until limited by the food supply, eaten by another, organism, destroyed in other ways. Some bacteria reproduce every 20 minutes. In more complex organisms, however, the life span of individual cells varies. "The governance of an organism such as a higher animal, which is a society of cells, dictates that some of the cells in the society will reproduce and that others will not," writes Mazia. "In general, the cells of tissues that perform special services for the entire organism, such as cells of the nervous system and the muscular system, do not reproduce at all." We are born with our full complement of brain neurons. Most of these neurons live as long as we do, but when one dies it is not replaced.

By contrast, the white phagocytic (scavenger) cells in our blood are replaced every two or three weeks. The life span of our red blood cells is three to four months. Our skin cells, too, keep being replaced. In such tissues, "The rate of production of new cells is nicely modulated to compensate for the continual loss of old cells," Mazia points out. In case of wounds or infections, when many cells are lost, the remaining cells multiply rapidly until the number of new cells is sufficient to replace what was lost and then stop. It is only in cancer that the cells repeat the cycle without restraint.

Surprisingly, all cells, whether of plant, bacteria, mouse or man, are made of the same fundamental materials: nucleic acids, proteins, lipids, carbohydrates, water and salts.

"The uniformity of the earth's life, more astonishing than its diversity, is accountable by the high probability that we derived, originally, from a single cell," notes Lewis Thomas. "It is from the progeny of this parent cell that we take our looks; we still share genes around and the resemblance of the enzymes of grasses to those of whales is a family resemblance."

The genetic material in all these cells is DNA (deoxyribonucleic acid), a large molecule which directs the making of duplicate cells according to a complex code. DNA also directs the building of proteins within the cell. Even the smallest and simplest living cells -- the mycoplasma -contain a relatively large amount of DNA, enough to code for up to a thousand different proteins. Every human cell has about 6 feet of DNA strands and every adult carries about 100 billion miles of ultra-thin DNA strands in his body -- a distance greater than the diameter of the solar system.

This DNA is concentrated in the cell's nuclear region. But there is a fundamental distinction between two major categories of cells:

The procaryotic cells, which include the bacteria, mycoplasma and blue-green algae, do not have any membrane around their nuclear region. In fact, they do not have any membrane-bound organelles at all.

The eucaryotic ("proper nucleus") cells do have a double membrane to separate their nucleus from the cytoplasm, as well as many other internal membranes to segregate their organelles. The cells of all animals and plants (except blue-green algae) and those of one-celled protozoa fall in this category. Only the eucaryotic cells are able to combine with one another to form multicellular systems -- an important step

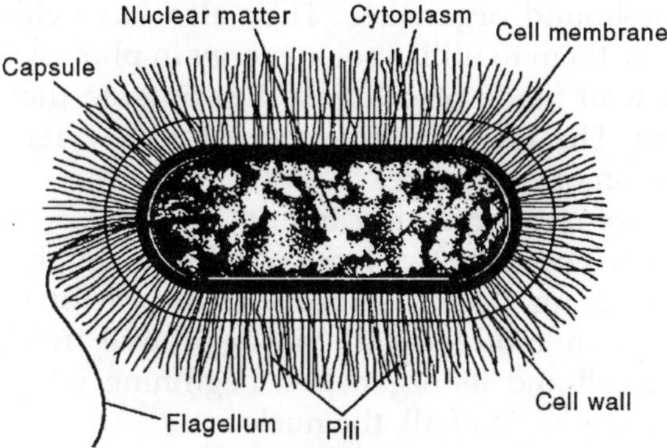

Fig. Bacterial Cells

Bacterial cells have no nucleus; their chromosomes, smaller and simpler than those of animal cells, are not enclosed in a membrane. As this diagram of an Escherichia coil bacterium shows, bacteria do not have any membrane -- bound organelles at all. They cannot combine with one another to form multicellular systems, as plant or animal cells do.

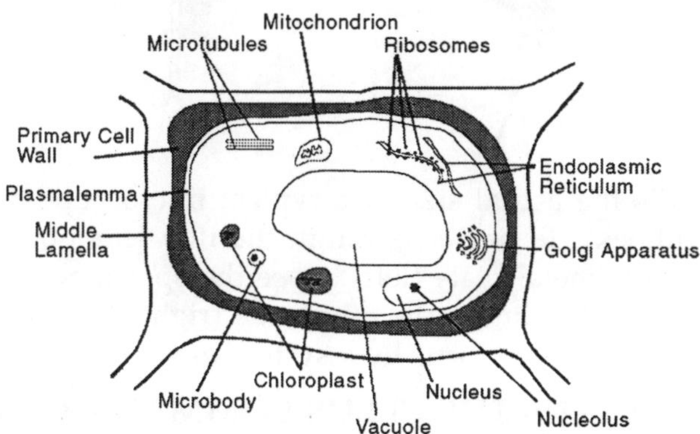

Fig. Plant Cell

And while, in general, the procaryotic cells produce only exact duplicates of themselves, the eucaryotic cells are capable of differentiation into many different kinds of cells, at least in higher organisms. This gives the eucaryotic cells certain obvious advantages. However, the procaryotes have advantages of one thousand times larger than bacterial cells, plant cells have a well-defined nucleus, as well as many membrane-bound organelles. They also have chloroplasts which allow them to utilize solar energy in photosynthesis. A rigid cell wall made of cellulose lies outside their plasma membrane. The large vacuole, containing water and salt, preserves osmotic pressure.

They are not necessarily inferior. As Mazia puts it, the differences between the two types of cells are simply "different ways of making a living." Some of the most important discoveries in modern cell biology have involved the eucaryotic cell and its organelles, beginning with the most prominent organelle of all, the nucleus.

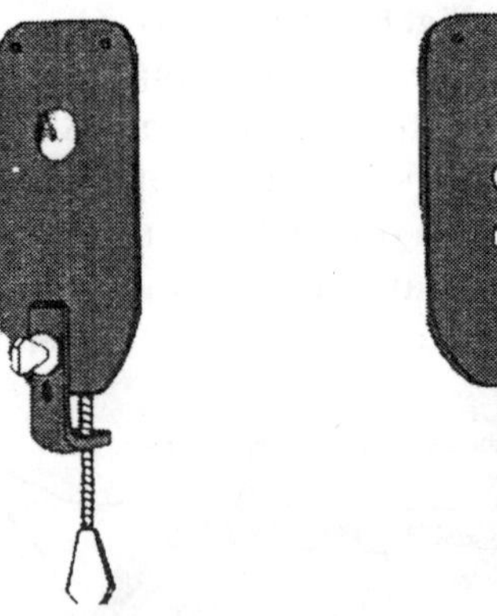

This is the actual size of a typical microscope built by Leeuwenhoek. He peered through the tiny lens opening on one side of a metal plate (left) to see the specimen mounted on the point of a pin on the other side (right). The specimen could be moved into focus by a system of screws.

THE NUCLEUS THE CELL'S COMMAND CENTRE

The nucleus is the biggest, densest, most obvious structure in the cell -- the first to be recognized by the microscopists and the first to be isolated in the biochemists' centrifuge.

Ironically, it was not a scientist but a cloth merchant, Anton van Leeuwenhoek, who saw the nucleus of a cell for the first time. Leeuwenhoek, who lived in Delft, Holland, ground his own lenses as a hobby -- once he made a lens out of a grain of sand. His hobby became a passion as he examined everything from pond water to the scum on his teeth. Unfortunately, he jealously guarded his methods and after he died it took more than a century for others to make equally good instruments. Even then they did not have his remarkable eyesight: He managed to perceive things through his single lens that modern cytologists need a compound microscope to see.

In 1702, while examining the blood cells of a fish, Leeuwenhoek discovered "a little clear sort of light in the middle," he reported in a letter to the Royal Society in London. This was the first inkling anyone had that animal cells were not just cavities, but had an internal structure. A century later, the nucleus of a plant cell was seen for the first time and in 1830 Johann Evangelista Purkinje, a Bohemian physiologist, described the relatively enormous nucleus of a hen's egg.

However, nobody knew what this nucleus did. Several researchers noted that before a cell divided, the nucleus divided. Yet it wasn't until the beginning of the 20th century that they grasped the connection between the rodlike chromosomes which had been observed in the nucleus and the transmission of hereditary traits. At that point, the importance of the nucleus became clear.

The nucleus is the cell's command centre. It contains the genes, the units of hereditary material (DNA) which give directions for everything the cell is and will be and thus controls the cell's reproduction and heredity. Yet it also responds to the environment. Only certain genes are "turned on" to give orders for the production of specific proteins at any particular time. The molecules which switch these genes on or off come from the cytoplasm, where they are generated as a result of interaction between the surface membrane and the environment. Thus the commands from the nucleus are always influenced by what goes on outside the cell. While the genes contain the total range of the cell's possibilities, the environment selects which of these possibilities for growth and development will be expressed.

Without a nucleus, cells die rapidly, excepting only the mammalian red blood cells -- the only common animal cells to be deprived of nuclei -- which do live a few months. Of course these cells have no progeny. (The mammalian red blood cells are formed in the bone marrow through a peculiar process in which half of the newly-made cells differentiate to become red blood cells, while the other half remain in the bone marrow, to produce a new crop of cells). Actually the mammalian red blood cells are just bags for hemoglobin, which carries oxygen through the body. Other cells whose nuclei have been removed surgically, or destroyed, survive only a few hours.

As might be expected, the nucleus is constantly active. Before cell reproduction, new DNA must be synthesized and every single gene must be replicated. These genes, which are linked together into chromosomes, are then separated into two duplicate sets. At this point the nuclear envelope breaks down, the two sets of chromosomes move to opposite poles of the cell and the cell divides, forming two identical daughter cells.

Meanwhile the nucleus goes through extraordinary changes in shape, from spherical to oval and to a strange assortment of folds or twists, until it actually disappears -- to be re-formed, in duplicate, at the end of cell division. In addition to all this reproductive activity, the genes in the nucleus selectively direct the synthesis of thousands of enzymes and other proteins in the cytoplasm. This is done through the "transcription" of information from various parts of the DNA molecules into new strands of nucleic acid (messenger-RNA) which carry it into the cytoplasm.

As long as the genes are busy replicating themselves or transcribing their information, the chromosomes into which they are linked cannot be distinguished -- they look like a jumble of threads. It is only during mitosis that these incredibly thin strands of DNA and protein become thick enough to be seen as rods under the light microscope. During this process the DNA strands condense through interaction with various proteins, coil themselves again and again and pack themselves into highly complex, tight bundles that are about 8,000 times shorter than the original strands.

It took years of effort and many technological advances before the chromosomes of all kinds of cells could be seen clearly enough to be counted accurately. Each species of plant or animal has a characteristic number of them and these chromosomes also vary in size, length and other properties. It was a major achievement when the number of chromosomes in human cells -- 46 -- was finally established in 1956 by Joe Hin Tjio of the NIH and Albert Levan of Sweden, who also succeeded in identifying each individual human chromosome by its characteristic shape.

The cells which will form a new human being -- the egg cells and sperm cells -- depart from this norm of 46 chromosomes per cell; each carries only half the usual complement. Thus, when a human egg and sperm fuse normally, the new cell again has a total of 46 chromosomes, or 23 pairs, with one chromosome in each pair coming from the mother and one from the father.

Several forms of mental retardation and dozens of other disorders have now been traced to gross errors in the number or shapes of the chromosomes in each cell. In recent years it has become possible to test for several of these prenatally, through a technique called amniocentesis. For example, Down's syndrome (mongolism) can now be detected before birth, by withdrawing a sample of amniotic fluid from the expectant mother's uterus with a hypodermic needle and counting the chromosomes in cells which the growing fetus shed in this fluid. (Down's syndrome is caused by an extra chromosome, believed to result from the incomplete separation of the chromosomes during the formation of the egg; this occurs most frequently when the mother is over 40 years old). Thousands of parents have taken advantage of this forecasting ability early in pregnancy.

The nuclei of individual cells are visible with a good light microscope at a magnification of only 100. At a magnification of 2,300, Tjio and Levan succeeded in examining individual chromosomes. But going one step further and investigating the genes, the units of heredity inside the chromosomes, proved far more difficult.

This is where the chemists and later the molecular biologists made some of their most important contributions. As early as 1869, a Swiss chemist, Fredrick Miescher, isolated some nuclear material from the pus cells on discarded hospital bandages and analyzed its content: phosphorus, carbon, oxygen, hydrogen and nitrogen. The material was later called nucleic acid and shown to consist of two kinds of substances, one containing the sugar ribose (this was ribonucleic acid, or RNA) and the other containing deoxyribose (this was deoxyribonucleic acid, or DNA). The substance in the chromosomes was identified as DNA. In 1944, Oswald Avery and his associates at the Rockefeller Institute discovered that DNA was directly involved in transferring hereditary characteristics from one strain of bacteria to another. And then everything speeded up.

By 1951, Maurice Wilkins and Rosalind Franklin at King's College, London, were studying the X-ray diffraction patterns of purified fibers of DNA -- patterns which proved crucial in understanding the structure of DNA. Two years later, in 1953, Francis Crick and James Watson, working at the Medical Research Council laboratories in Cambridge, England, proposed their famous double helix model of the DNA molecule, which explained how DNA is built and how it replicates. Watson, Crick and Wilkins later won the Nobel Prize for this achievement.

From then on, genes -- the units of heredity whose existence was only hypothesized by earlier scientists -- could be studied and analyzed in terms of their biochemical composition. According to Watson and Crick, the DNA molecule consists of different sequences of nucleotides (DNA building blocks linked together in helical or spiral form, like an almost endless, twisted rope ladder). A single gene might be a section of this rope ladder perhaps 2,000 steps long. And every step, or rung, in the ladder is made up of two nucleotides which fit together according to their bases. (Nucieotides consist of one of four chemical bases -- adenine, thymine, guanine and cytosine -- attached to a chain of sugar-phosphates which acts as an outer backbone.

An adenine base (A) will fit only with thymine (T) and a guanine base (G) will fit only with cytosine (C), so that the sequence of bases on one half of the rung (for example, AGCG) determines the sequence on the other half (TCGC). This is the alphabet of the nucleic acids -- a small set of "letters" with which, as with the ABC's, an infinite number of messages and instructions can be written.

In the early 1960's, the language in which such instructions were written -- the genetic code -- was deciphered by Marshall Nirenberg of NIH, Severo Ochoa of New York University and Har Gobind Khorana of the University of Wisconsin, for which they, too, won the Nobel Prize. The code consists of triplets of nucleotides which are "read" in sequence along the DNA molecule. Each triplet corresponds to one word, i.e., one of the 20 amino acids, which are the building blocks of protein. This is the universal language of life

Each gene is a series of triplets which gives the instructions for building a specific protein and thus influences a specific trait, such as blue eyes. In 1976, after nine years of work, Khorana produced an artificial gene out of laboratory chemicals and showed that it functioned normally when he inserted it in a live bacterium. Meanwhile, a new technique using "recombinant DNA" was developed for studying individual genes: After splitting some animal or human DNA into thousands of segments, scientists transfer one segment into a bacterium, where it is replicated and expressed together with the bacterium's own genetic material.

In many ways, then, scientists are developing the ability to manipulate genes. They are also learning to mitigate or prevent some of nature's genetic mistakes. Genetic diseases used to be considered quite rare, but it is now recognized that innumerable persons suffer the consequences of disorders due wholly or in part to a defective gene or chromosome. Some 2,000 different genetic disorders have been identified.

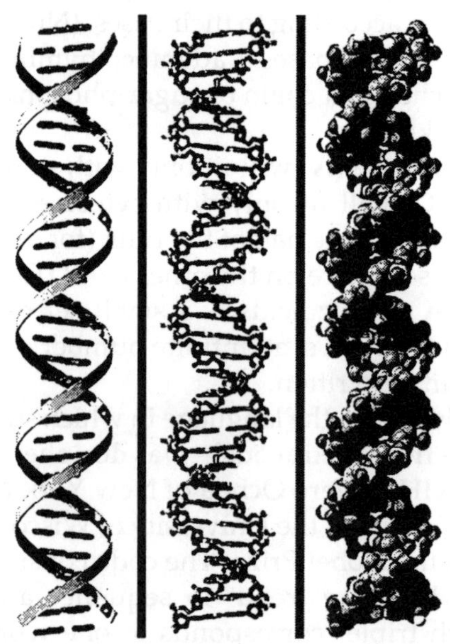

Fig. The Double Helix: Model of the DNA (Deoxyribonucleic Acid) Molecule which Embodies the Code of Heredity.

First Letter	Second Letter: T	Second Letter: C	Second Letter: A	Second Letter: G	Third Letter
T	TTT, TTC } Phe TTA, TTG } Leu	TCT, TCC, TCA, TCG } Ser	TAT, TAC } Tyr TAA Stop TAG Stop	TGT, TGC } Cys TGA Stop TGG Trp	T C A G
C	CTT, CTC, CTA, CTG } Leu	CCT, CCC, CCA, CCG } Pro	CAT, CAC } His CAA, CAG } Gln	CGT, CGC, CGA, CGG } Arg	T C A G
A	ATT, ATC, ATA } Ile ATG Met	ACT, ACC, ACA, ACG } Thr	AAT, AAC } Asn AAA, AAG } Lys	AGT, AGC } Ser AGA, AGG } Arg	T C A G
G	GTT, GTC, GTA, GTG } Val	GCT, GCC, GCA, GCG } Ala	GAT, GAC } Asp GAA, GAG } Glu	GGT, GGC, GGA, GGG } Gly	T C A G

Fig. The Genetic Code

Sickle-cell disease, for example -- a disease which affects some 50,000 Americans, mostly blacks, producing chronic anemia, jaundice, severe pain, poor resistance to infection and sometimes an early death -- has been traced to a single gene and the resulting misplacement of just one amino acid out of 300 in the hemoglobin molecules of the victims' red blood cells. The cells then become distorted from their normal round shape, often into the shape of a crescent or sickle and are altered in other ways as well. The abnormal cells may stick together, obstructing the small blood vessels and causing damage as well as pain, or be removed too rapidly by the spleen, causing anemia.

Over 2 million Americans are carriers of this defect. Although they may be perfectly healthy, if they marry one another they risk transmitting sickle-cell disease to their offspring: Each of their children will have a 25 per cent chance of inheriting the disease. A simple blood test can now identify people who carry the sickle-cell gene. Researchers are studying various techniques for detecting the defect prenatally and several possible methods of treatment are under investigation.

While genetic defects cannot be corrected inside the cell -- at least not with present techniques -- increasing knowledge about their specific effects on human development has led to various forms of environmental treatment: medications, diet, or a change in life habits.

Eventually, perhaps, it may become possible to turn on specific genes in the cell at specific times. Although every cell in the body has the same set of genes and all of these genes must be replicated before cell division, only certain genes are turned on to produce specific proteins at any given time. In specialized cells such as liver cells or bone cells, many genes are turned off permanently.

But sometimes there are problems. In sickle-cell disease, for instance, the gene which makes normal hemoglobin in the fetus gets turned off after birth -- as it should -- but the newly activated, adult gene is defective. As scientists learn more about the switching-on and switching-off mechanisms, they may be able to avert the switch-off in such cases, or else re-activate the embryonic gene, which might lead to a treatment for the disease. Looking even farther into the future, researchers may yet learn to insert missing genes right into human cells to correct certain genetic mistakes.

RIBOSOMES, OUR PROTEIN FACTORIES

Whenever the cell needs a protein, the cytoplasm alerts the nucleus with an as-yet-unidentified chemical message. The DNA of the appropriate gene or genes in the nucleus then issues the specific instructions for making this protein.

These instructions are transmitted not by the DNA itself, but by a close copy made out of a different nucleic acid, RNA. The original DNA remains safely in the nucleus, somewhat like the printing block in a printing press. The RNA copy is manufactured in the nucleus by transcribing just one chain of the DNA double helix (one side of the twisted ladder), which is enough to code the instructions. After additional processing, this messengerRNA then crosses the nuclear membrane and goes out into the cytoplasm, where its instructions will be carried out by tiny organelles called ribosomes, the "factories" in which the protein molecules are made.

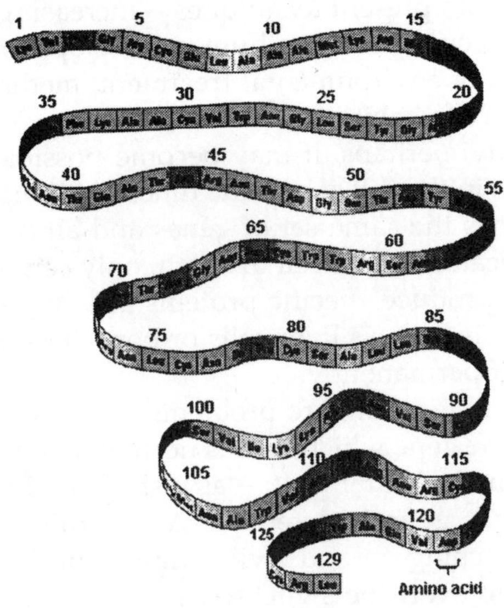

Fig. Primary Structure of Lysozyme

The ribosomes really are tiny: about 200 angstroms in

diameter, or less than one millionth of an inch, so small that their internal structure cannot be seen even under an electron microscope. However, most cells contain thousands and sometimes even millions of them. Up to 30 separate ribosomes may be attached to a single molecule of messenger-RNA, each ribosome making its own protein as it works along the chain. Within these ribosomes, free amino acids are joined together in accordance with the sequences specified by the messenger-RNA.

The primary structure of lysozyme, one of the smallest known enzymes, is this linear sequence of 129 amino acids. The primary structure determines the coiling of the peptide chains into folds and spirals (the secondary structure) and into a complex 3-D structure (the tertiary structure). The tertiary structure brings together amino acids which might otherwise be at opposite ends of the chain.

The active site of lysozyme is shown as a crevice in this 3-D model. Lysozyme digests the cell walls of bacteria that are found in nasal mucus and other secretions. Fitting together like a key in a lock, a piece of carbohydrate from a bacterial cell wall binds to the lysozyme's active site, where it is split by amino acids on either side.

The linear sequence of amino acids in a protein is known as the protein's primary structure. However, proteins never exist as straight chains -- they are always coiled -- and they often contain several different but interconnected chains of polypeptides (linked amino acids). Regions of these chains coil into spirals which give the protein its secondary structure. As a result of interactions of the various side chains, the molecule as a whole then folds and coils further into a complex 3-dimensional structure, the protein's tertiary structure. The primary structure determines the secondary and tertiary structures. The tertiary structure controls function. When all the required amino acids for a specific protein are joined in a completed chain within the ribosomes, the chain is released as a free protein -- and coils up into a characteristic 3-dimensional shape, on which its activity depends.

Despite their small size, ribosomes make up a large part

of cells in which protein is manufactured. In the Escherichia coli bacteria, for instance, they account for about one-fourth of the total cell mass; this great number of ribosomes allows many proteins to be made simultaneously in one cell. It takes about a minute or two for a protein to be completed.

Ribosomes were first discovered in the middle 1950's, when biochemists became aware of these particles' role in protein synthesis at the same time that George Palade, working with an electron microscope, saw them in the cytoplasm of cells and described their gross structure. Shortly afterwards, Palade and Philip Siekevitz of Rockefeller University and other investigators isolated these particles, which were later called ribosomes.

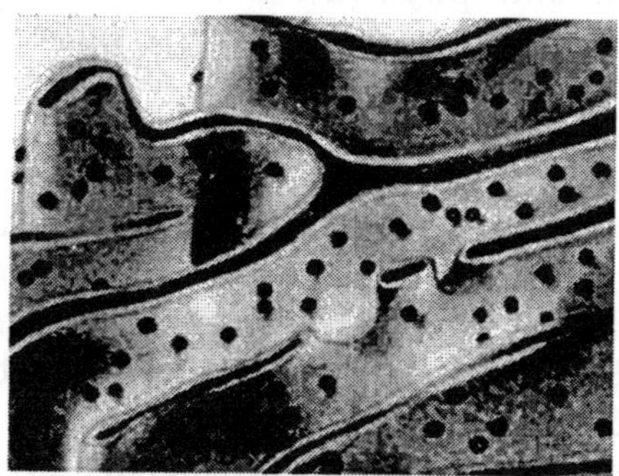

Fig. The Endoplasmic Reticulum

By now it is known that ribosomes are made of two unequal sub-units and that each of these consists of RNA and proteins. This RNA (ribosomal-RNA) differs from messenger-RNA; it is produced in the nucleolus, a prominent, globular structure in the nucleus. (A third type of RNA, transfer-RNA, is also involved in protein synthesis, particularly in bringing to the ribosomes the amino acids specified by the messenger-RNA). The ribosomes of bacteria are somewhat simpler than those of eucaryotic cells. Recently Masayasu Nomura of the University of Wisconsin succeeded in taking apart the RNA and proteins in sub-units of bacterial ribosomes and then putting these components back together in the proper order, so that the sub-units functioned normally in a test tube. This was an important step towards understanding how the more complex ribosomes in human cells are made -- and what happens when they malfunction.

In general, ribosomes fall into two categories: Those that are free in the cytoplasm and those that are bound to membranes. The two kinds of ribosomes play similar roles in the manufacture of proteins. But while the free ribosomes apparently leave the proteins equally free to float in the cytoplasm, the bound ribosomes transfer their finished proteins into a gigantic, cobwebby organelle made up of membranes, the endoplasmic reticulum. The endoplasmic reticulum.

Index

N

O

P

R

S

T

W

X